HOW TO USE A HAM RADIO

Navigating Frequencies: The Essential Guide for Amateur Radio Enthusiasts

Fitzpatrick J. Thompkins

Copyright © 2024 by **Fitzpatrick J. Thompkins**

All rights reserved

No part of this publication may be reproduced, stored in a retrieval system, or transmitted, in any form or by any means, electronic, mechanical, photocopying, recording, or otherwise, without the prior written permission of the author.

The information in this ebook is true and complete to the best of our knowledge. All recommendation are made without guarantee on the part of author or publisher. The author and publisher disclaim any liability in connection with the use of this information.

Table of Contents

Introduction — 5
- Overview of Ham Radio — 7
- Importance of Amateur Radio in Today's World — 9
- Basic Concepts and Terminology — 11

Chapter: 1 Getting Licensed — 13
- Understanding Different Classes of Ham Radio Licenses — 13
- Study Tips for the Licensing Exam — 15
- How to Apply for Your License and Call Sign — 17

Chapter: 2 Setting Up Your First Ham Radio Station — 19
- Choosing the Right Equipment — 19
- Basic Setup Guide — 21
- Safety Considerations — 23

Chapter: 3 Operating Basics — 25
- Understanding Frequencies and Bands — 25
- How to Tune Your Radio — 27
- Making Your First Contact — 29
- Q Codes and Radio Etiquette — 31

Chapter: 4 Exploring Ham Radio Activities — 33
- Contesting — 33
- DXing (Long Distance Communication) — 35
- Digital Modes of Communication — 37
- Emergency Communications and Public Service — 39

Chapter: 5 Antennas and Propagation   **41**
   Types of Antennas   41
   Basic Antenna Theory   43
   Understanding Radio Wave Propagation   45

Chapter: 6 Advanced Operation Techniques   **47**
   Adjusting Your Radio for Optimal Performance   47
   Using Amplifiers for Increased Power   49
   Advanced Digital and Satellite Communications   51

Chapter: 7 Joining the Ham Radio Community   **53**
   Finding and Joining Local Clubs   53
   Participating in Nets and Roundtables   55
   Ham Radio Conferences and Events   57

Chapter: 8 Maintenance and Troubleshooting   **59**
   Routine Maintenance Checks   59
   Common Issues and How to Resolve Them   61
   Upgrading Your Equipment   63

Chapter: 9 The Future of Ham Radio   **65**
   Technological Advances and Trends   65
   The Role of Ham Radio in Modern Communications   67
   Continuing Your Education in Radio Technology   69

Conclusion   **71**

Introduction

In the quiet suburb of Pineview, nestled between the bustle of the city and the tranquility of the countryside, lived Alex Carter. Alex, a retired teacher with a spark of curiosity that age couldn't dull, had always been fascinated by the mysteries of communication. His interest led him to a local flea market, where he stumbled upon a book that seemed to whisper promises of adventure: "How to Use a Ham Radio."

With the book tucked under his arm, Alex returned home, feeling as if he'd found a treasure map. He poured over the pages that evening, each chapter unfurling like a roadmap to a world he had never explored. The book wasn't just a manual; it was a call to adventure, offering a journey into the heart of communication, community, and discovery.

The first chapters laid the foundation, explaining the essence of ham radio and its significance in the modern world, from providing emergency communications during disasters to connecting with people across the globe. It spoke of a community of enthusiasts, ready to welcome newcomers with open arms. Alex felt a stir of excitement. The book wasn't just selling him a hobby; it was inviting him to become part of a worldwide fellowship.

As Alex delved deeper, the book guided him through the process of getting licensed, choosing the right equipment, and setting up his first station. It demystified complex concepts with ease, making them accessible and engaging. Alex appreciated how the book emphasized the importance of proper licensing and ethical operation, reinforcing the sense of responsibility that came with wielding such a powerful tool of communication.

The chapters on operating basics and exploring ham radio activities were where the book truly shone. It presented a plethora of activities, from contesting and DXing to emergency communications, each with its allure. The book promised that with ham radio, Alex could chase distant signals, contribute to his community's safety, or simply enjoy a casual chat with someone from a culture he'd only read about.

Maintenance, troubleshooting, and the future of ham radio were covered with equal thoroughness, preparing Alex for the practical aspects of his new hobby and exciting him with glimpses into its evolving landscape. Appendices offered a glossary and recommended resources, ensuring that help was always at hand.

But it was the conclusion that sealed Alex's resolve. It spoke of ham radio not just as a hobby, but as a lifelong journey of learning, exploration, and community. It reminded him that every transmission sent out into the ether was a testament to human connection and curiosity.

Inspired, Alex decided to embark on this journey. Months later, as he made his first contact with a fellow ham halfway across the world, he realized that the book had delivered on its promise. It hadn't just taught him how to use a ham radio; it had opened a door to a world where every frequency held a story, every call sign was a potential friend, and every transmission bridged hearts and minds across the vastness of our planet.

For those standing at the threshold of this expansive world, wondering whether to step in, "How to Use a Ham Radio" is more than a guide. It's an invitation to join a global adventure that transcends borders, cultures, and generations. Alex would tell you, it's an invitation worth accepting.

Overview of Ham Radio

Ham radio, also known as amateur radio, is a fascinating and multifaceted hobby that connects people from various parts of the globe, transcending geographical boundaries and cultural differences. At its core, ham radio involves the use of radio frequency spectrum for purposes of non-commercial exchange of messages, wireless experimentation, self-training, private recreation, radiosport, contesting, and emergency communication. The term "ham" as applied to 1908 was the call letters of one of the first amateur wireless stations operated by some of the amateurs of the Harvard Radio Club.

One of the unique aspects of ham radio is its ability to function independently of the internet or mobile phone networks, making it an invaluable tool during natural disasters or other emergencies when traditional communication channels fail. Ham radio operators, known as hams, play a crucial role in disaster relief efforts by providing essential communication services that aid in rescue and recovery operations.

Getting started with ham radio requires obtaining a license, which ensures that operators understand the regulations and technical knowledge necessary to use the equipment safely and effectively. The licensing process involves passing an examination that covers a range of topics, including radio theory, regulations, and operating practices. Once licensed, hams are assigned call signs, which serve as their unique identifier on the airwaves.

The world of ham radio offers a diverse range of activities to explore. From making casual contacts with other hams around the world, known as rag chewing, to participating in contests or radiosport, which challenge operators to make as many contacts as possible within a specific time frame. Digital modes of communication, such as FT8 and PSK31, allow hams to communicate using computer-generated modes, expanding the possibilities for experimentation and contact.

Ham radio also serves as a platform for technical innovation and experimentation. Many hams build their own radios and antennas, experimenting with new technologies and contributing to advancements in the field. Satellite communication is another exciting aspect of ham radio, allowing operators to communicate via amateur radio satellites orbiting the Earth, further extending the reach of their communications.

In addition to the technical and operational aspects of ham radio, the hobby fosters a sense of community among its practitioners. Ham radio clubs and organizations offer opportunities for hams to meet, share knowledge, and participate in group activities. These communities not only support learning and development in the hobby but also form a global network of individuals united by their passion for amateur radio.

Ham radio's appeal lies in its blend of technology, communication, and community. Whether it's the thrill of making a contact with a distant country, the satisfaction of building and operating one's own equipment, or the fulfillment of providing critical communication services during emergencies, ham radio offers a wide range of experiences to explore. For those looking to embark on this exciting journey, understanding the overview of ham radio is just the beginning. With a license in hand and a spirit of curiosity, the airwaves open up a world of discovery and connection, echoing the timeless human desire to communicate and connect.

Importance of Amateur Radio in Today's World

Amateur radio, often referred to as ham radio, remains a vital and thriving form of communication in today's world, despite the rapid advancements in technology and the proliferation of digital communication platforms. Its importance spans several facets, from emergency preparedness to fostering global connections and supporting educational endeavors.

In the realm of emergency preparedness, amateur radio stands out as a reliable means of communication during natural disasters and other emergencies when conventional communication infrastructures are compromised. Ham radio operators have historically played pivotal roles in providing critical communication links during hurricanes, earthquakes, and other catastrophic events. Their ability to operate independently of the internet and cellular networks makes ham radio an invaluable tool for emergency responders and disaster relief organizations.

Beyond its utility in emergencies, amateur radio also serves as a bridge between cultures and nations. It transcends geographical boundaries, enabling individuals from all corners of the globe to connect, share experiences, and foster mutual understanding. These person-to-person connections contribute to a sense of global community and can break down barriers that often divide people.

Furthermore, amateur radio is a powerful educational tool. It offers a hands-on learning experience for individuals interested in science, technology, engineering, and mathematics (STEM) fields. Through the process of setting up and operating their own radio stations, enthusiasts gain practical knowledge in electronics, physics, and radio wave propagation. Schools and educational organizations often incorporate ham radio into their curriculum to stimulate interest in STEM subjects and encourage critical thinking and problem-solving skills.

The hobby also promotes innovation and experimentation. Ham radio operators are known for their ingenuity, often building and modifying their own equipment and experimenting with different modes of communication. This spirit of innovation not only advances the field of radio communication but also contributes to technological developments that can have broader applications.

In addition to its practical applications, amateur radio is a hobby that offers personal growth opportunities. It requires operators to obtain a license, which involves learning technical information and passing an examination. The process of becoming a licensed ham radio operator instills a sense of achievement and encourages lifelong learning. Moreover, the community aspect of amateur radio provides a supportive environment for individuals to develop communication and leadership skills.

Amateur radio also plays a role in environmental monitoring. Many ham radio enthusiasts engage in activities like tracking weather balloons and satellites, contributing valuable data for weather forecasting and scientific research. This involvement not only aids in environmental conservation efforts but also offers amateurs a chance to participate in cutting-edge science projects.

Given its multifaceted contributions to emergency communications, global connectivity, education, innovation, personal development, and environmental monitoring, amateur radio continues to be a significant and relevant hobby in today's world. Learning how to use a ham radio not only opens up a world of communication possibilities but also aligns with broader societal needs and values. It's a hobby that offers something for everyone, whether they're drawn to the thrill of emergency communications, the joy of international friendship, the challenge of technical innovation, or the satisfaction of personal achievement.

Basic Concepts and Terminology

In exploring the world of ham radio, understanding the basic concepts and terminology is like learning the language of a new country. It allows for better navigation, deeper understanding, and more meaningful interactions within the amateur radio community. At the heart of ham radio are the frequencies and bands, the invisible highways of airwaves that hams use to communicate. These are segmented into specific ranges, each with its own characteristics and rules, enabling a variety of communications from local chatter to global messages.

Central to the operation of a ham radio is the concept of a transceiver, a device that combines both a transmitter and a receiver in one unit. This pivotal piece of equipment allows hams to send out their own signals and listen to others, facilitating a two-way conversation that can span continents. Antennas are another crucial component, serving as the bridge between the radio waves and the equipment. Their designs and placements are varied, each tailored to specific frequencies and purposes, whether it's a simple wire strung between trees for local nets or a sophisticated beam antenna pointing towards distant horizons for international contacts.

Power sources for ham radios also vary, from handheld devices with battery packs to full station setups that require a steady and reliable power supply. Understanding the power requirements and limitations is essential for both effective communication and safety.

Modulation modes are the methods by which information is encoded onto radio waves. These include voice modes like FM (Frequency Modulation) and SSB (Single Side Band), as well as digital modes that use computers to encode, send, and decode messages. Each mode has its advantages and niches within the ham radio spectrum, from the clarity of FM in local communications to the efficiency of digital modes in low-signal conditions.

Call signs are unique identifiers assigned to licensed operators, acting as both a greeting and a signature in the ham radio world. These are issued by regulatory bodies and follow a specific format that can often hint at the operator's geographic location.

Q codes and phonetic alphabets are shorthand languages used in ham radio to convey common questions or statements in a universally understood form. Q codes originated from telegraphy, offering a concise way to express complex questions or instructions, while the phonetic alphabet ensures clarity when spelling out call signs, especially in challenging reception conditions.

Understanding propagation, the study of how radio waves travel and interact with the environment, is key to mastering ham radio communications. Factors like time of day, solar activity, and atmospheric conditions can dramatically affect the reach and clarity of transmissions. Hams use this knowledge to predict the best times and frequencies for making long-distance contacts.

Finally, the legal and ethical aspects of ham radio cannot be understated. Regulations governing frequencies, power levels, and permissible content ensure that the airwaves remain open and accessible to all users. Etiquette, or the unwritten rules of conduct, fosters a culture of respect, assistance, and camaraderie within the community.

In summary, the rich tapestry of ham radio is woven from its frequencies, equipment, operating modes, and the language that binds its community. Understanding these basics not only equips newcomers with the tools they need to start their journey but also deepens the appreciation for the art and science of amateur radio. With this foundation, the world of ham radio opens up as a field of endless exploration and connection.

Chapter: 1 Getting Licensed

Understanding Different Classes of Ham Radio Licenses

Understanding the different classes of ham radio licenses is a crucial step for anyone looking to dive into the world of amateur radio. In the United States, the Federal Communications Commission (FCC) oversees the licensing process, offering three distinct classes of licenses, each with its unique set of privileges and requirements. These licenses are the Technician Class, the General Class, and the Amateur Extra Class. As you move up in license class, you gain access to more frequencies and operating rights.

1. Technician Class License

The Technician Class license is the entry-level license. It's designed for beginners and gives access to all Amateur Radio frequencies above 30 MHz, allowing for communication locally and within North America. It also includes limited privileges on the HF (High Frequency) bands, which are used for long-distance communication. To obtain this license, you must pass a 35-question multiple-choice examination covering basic regulations, operating practices, and electronics theory.

How to Use a Ham Radio as a Technician:
- VHF/UHF Operations: Most new hams start with 2-meter and 70-centimeter bands. These are ideal for local communications through repeaters.
- Digital Modes: Technicians can also operate digital modes on bands above 30 MHz, which is great for digital voice and data communications.
- Limited HF Access: Use the limited HF privileges for making longer-distance contacts, particularly on 10 meters, where Technician operators have CW (Morse Code), digital, and phone privileges.

2. General Class License

The General Class license opens up significantly more spectrum on the HF bands, which are ideal for international communications. To upgrade to General, you must pass a 35-question exam that covers more advanced topics in radio theory, regulations, and operating practices.

How to Use a Ham Radio as a General:
- HF Bands: Explore the world by making contacts on the HF bands. General licensees have access to most of the HF spectrum, allowing for global communication.
- Advanced Modes: With more bands available, you can experiment with various modes like RTTY (Radioteletype),

PSK31 (a digital mode), and SSB (Single Side Band) voice communication.
- Emergency Communication: General operators can participate more fully in emergency communication efforts and nets that operate on the HF bands.

3. Amateur Extra Class License

The Amateur Extra Class represents the highest level of licensing, granting all available U.S. Amateur Radio operating privileges on all bands and all modes. Passing the 50-question exam requires a deeper understanding of radio theory, electronics, and FCC regulations.

How to Use a Ham Radio as an Amateur Extra:
- Full HF Access: Enjoy the privilege of operating across all amateur bands. This includes exclusive frequency allocations.
- Leadership Roles: Many clubs and groups require an Amateur Extra license for leadership positions.
- Mentoring: With the highest level of knowledge and operating privileges, Amateur Extra operators are well-equipped to mentor newcomers and contribute to advancing the amateur radio hobby.

Getting Licensed

The journey to becoming a licensed ham operator begins with education and study. Numerous resources are available, including study guides, online courses, and local amateur radio clubs that offer classes and examination sessions. Once you're ready, you can find an examination session near you through the American Radio Relay League (ARRL) website or other amateur radio organizations.

Conclusion

Ham radio is a hobby that connects people across the globe, offers a vital communication link during emergencies, and fosters a community of individuals passionate about radio technology. By understanding and obtaining the appropriate class of license, you're unlocking the door to a world of communication possibilities, continuous learning, and technical exploration. Whether you're aiming to communicate locally on the VHF/UHF bands as a Technician, exploring the vast HF spectrum as a General, or enjoying the full privileges of an Amateur Extra, each step in your ham radio journey offers new opportunities and challenges.

Study Tips for the Licensing Exam

Studying for a ham radio licensing exam in the United States is an exciting step towards joining the amateur radio community. The process involves understanding the technical aspects of ham radio operation, regulations, and safety practices. Here are detailed study tips to help you prepare for the licensing exam and become a licensed amateur radio operator:

1. Understand the Exam Structure

First, familiarize yourself with the structure of the exam. The Federal Communications Commission (FCC) offers three levels of licenses: Technician, General, and Extra. Each level grants different privileges and requires passing a progressively more challenging exam. Knowing the structure and content scope of the exam you're preparing for will guide your study strategy.

2. Obtain Study Materials

Several reputable sources offer study guides and materials tailored to the licensing exams. These include:
- ARRL (American Radio Relay League) offers study guides for all license classes.
- Gordon West study guides, which are popular for their detailed explanations and practice questions.

- Online courses and video tutorials that cover the exam material in an interactive format.

3. Use Practice Exams

Taking practice exams is crucial for success. They familiarize you with the exam format and types of questions asked. Free online resources like HamStudy.org offer practice exams for all license classes and track your progress over time.

4. Join a Study Group

Studying with others can enhance your learning experience. Join local or online amateur radio clubs where members often organize study sessions for upcoming exams. These groups provide support, clarification of difficult concepts, and practical experience.

5. Focus on Weak Areas

As you study and take practice exams, identify areas where you're struggling. Allocate more study time to these topics to ensure a well-rounded understanding of the exam material.

6. Understand FCC Rules and Regulations

A significant portion of the exam covers FCC rules and regulations governing amateur radio. Make sure to study these thoroughly, as understanding legal aspects is crucial for responsible ham radio operation.

7. Get Hands-On Experience

If possible, gain practical experience with ham radio equipment. This could be through a local amateur radio club, mentoring from a licensed amateur, or even online simulations. Hands-on experience will deepen your understanding of how radio theory applies in real-world situations.

8. Review Technical Concepts

Especially for the General and Extra license exams, a solid understanding of electronics and radio theory is essential. Review technical concepts such as circuit diagrams, frequency allocations, antenna theory, and signal propagation.

9. Develop a Study Schedule

Create a realistic study schedule leading up to your exam date. Consistent, focused study sessions are more effective than cramming all the material at the last minute.

10. Stay Motivated

Remind yourself why you're pursuing a ham radio license. Whether it's for emergency communication, community service, or exploring radio technology, keeping your goals in mind will motivate you through the study process.

Final Thoughts

Preparing for a ham radio licensing exam takes time and effort, but it's a rewarding process that opens up a world of communication and community. By understanding the exam structure, using quality study materials, practicing with exams, and getting involved with the amateur radio community, you're setting yourself up for success. Remember, the journey doesn't end with getting licensed; it's just the beginning of your adventures in amateur radio.

How to Apply for Your License and Call Sign

Applying for your license and receiving a call sign are essential steps in becoming a licensed amateur radio operator, commonly known as a "ham." This process enables you to legally use ham radios, which are powerful communication tools for both hobbyists and professionals. Here's a comprehensive guide on how to get licensed to use a ham radio in the United States.

Understanding Ham Radio Licensing

In the U.S., the Federal Communications Commission (FCC) regulates amateur radio licenses. There are three classes of licenses: Technician, General, and Amateur Extra. Each offers increasing levels of access to radio frequencies and requires passing a written exam.

Step 1: Study for the Exam

First, choose which class of license suits your needs. Most beginners start with the Technician class. There are various resources available for studying:

- Books and Study Guides: Look for updated materials that cover the latest question pool.
- Online Courses and Practice Exams: Websites like the American Radio Relay League (ARRL) offer courses and practice tests.

Step 2: Find an Exam Session

Once you're ready, find an FCC-authorized Volunteer Examiner Coordinator (VEC) session. The ARRL website is a great resource for finding exam locations and dates near you.

Step 3: Take the Exam

- What to Bring: A government-issued photo ID, your Social Security Number (SSN) or FCC Registration Number (FRN), and any applicable exam fees.
- During the Exam: The tests are multiple-choice. The Technician and General exams have 35 questions, and the Amateur Extra has 50.

Step 4: Obtain Your FRN

Before taking the exam, you should obtain an FCC Registration Number (FRN). Register on the FCC's Commission Registration System (CORES) website. This number is used in place of your SSN for all transactions with the FCC.

Step 5: Pass the Exam and Wait for Your Call Sign

After passing the exam, your VEC will submit the results to the FCC. It usually takes a few days to a couple of weeks for the FCC to process your application and issue your license and call sign. You can check the status of your license on the FCC's Universal Licensing System (ULS) database.

Step 6: Getting on the Air

Once you have your call sign, you're legally allowed to operate on the amateur radio frequencies designated for your license class. Here are a few tips for getting started:

- Join a Club: Local amateur radio clubs are excellent resources for new hams. They can help with equipment, operating techniques, and making your first contacts.
- Choosing Equipment: Start with a VHF/UHF transceiver to access local repeaters or an HF transceiver for longer-distance communication, depending on your license class and interests.
- Learn Operating Procedures: Familiarize yourself with the etiquette and protocols of ham radio communication. Listening to experienced operators can be incredibly educational.

Conclusion

Becoming a licensed amateur radio operator opens up a world of communication possibilities. Whether you're interested in local community service, emergency response, or global communication, the ham radio community is welcoming and eager to help beginners. Follow these steps, and you'll be on your way to exploring the airwaves in no time.

Chapter: 2 Setting Up Your First Ham Radio Station

Choosing the Right Equipment

Choosing the right equipment is a crucial step in setting up your first ham radio station. Ham radio, or amateur radio, is a popular hobby and service that brings people, electronics, and communication together. People use ham radio to talk across town, around the world, or even into space without the Internet or mobile phones. It's a fun way to explore the world of communications technology, but getting started requires some basic knowledge of the equipment needed: transceivers, antennas, and power sources. Let's delve into how to select the appropriate gear for your ham radio setup.

Transceivers

A transceiver is a device that can both transmit and receive communications, making it the heart of your ham radio station. When choosing a transceiver, consider the following:

- Type: There are handheld, mobile, and base station transceivers. Handhelds are portable but have limited range. Mobile units are designed for vehicle installation with better range, and base

stations are intended for home use with the highest power and range.
- Frequency Bands: Ensure the transceiver covers the frequency bands you intend to use. Most beginners start with the 2-meter band (144-148 MHz) or the 70-centimeter band (420-450 MHz).
- Modes: Look for a transceiver that supports multiple modes of operation, such as FM (Frequency Modulation), SSB (Single Side Band), and digital modes. This versatility will allow you to explore different aspects of ham radio.
- Features: Consider ease of use, display quality, built-in features like automatic tuners, and connectivity options for computers or other digital devices.

Antennas

The antenna significantly affects your station's ability to send and receive signals. Your choice will depend on your location, the space available, and the frequencies you plan to operate on:

- For VHF/UHF: A simple vertical antenna or a Yagi antenna can be great starting points. Verticals are omnidirectional and easy to install, while Yagis are directional and offer greater range.
- For HF Bands: Options include dipoles, verticals, or wire antennas. Dipoles are versatile and can be installed in various configurations. Verticals require less space but might need a good ground system to perform well.

- Installation: Consider your installation options. Antennas can be mounted on roofs, balconies, or even inside attics. Make sure your antenna has a clear view of the sky for the best performance.

Power Sources

Reliable power is critical for operating your ham radio station:

- For Handhelds: Most come with rechargeable batteries. It's wise to have a spare battery or a method to charge the battery during long operations.
- For Mobile and Base Stations: These typically run on 12V power. A stable power supply that can convert household AC to 12V DC is essential for base stations. Mobile units can be powered directly from a vehicle's electrical system.
- Backup Power: Consider investing in a UPS (Uninterruptible Power Supply) or alternative power sources like solar panels or generators for emergency situations or to enjoy radio operation during field days.

Setting Up and Using Your Ham Radio

Once you have your equipment, setting up your station involves connecting your transceiver to your antenna and power source. Carefully read the manuals for each piece of equipment to understand the setup process and safety precautions.

- Tuning Your Antenna: Use an SWR (Standing Wave Ratio) meter to ensure your antenna is properly tuned to the frequencies you intend to use. A well-tuned antenna will transmit and receive signals more effectively and protect your transceiver from damage.
- Join a Club: Many ham radio clubs offer classes and mentoring for newcomers. Joining a club can accelerate your learning and provide valuable assistance in setting up your station.
- Practice: Start by listening to broadcasts and getting accustomed to the radio frequencies. Practice transmitting with a local repeater or by checking into a net to become comfortable with making contacts.

Conclusion

Setting up your first ham radio station is an exciting journey into the world of amateur radio. By carefully selecting your transceiver, antenna, and power source, you can build a station that meets your communication needs and interests. Remember, ham radio is not only about the equipment but also about the global community of enthusiasts ready to share their knowledge and experiences. Welcome to the world of ham radio!

Basic Setup Guide

Setting up your first ham radio station can be an exciting venture into the world of amateur radio. This comprehensive guide aims to streamline the process, ensuring you have a smooth and enjoyable start. Ham radio, also known as amateur radio, allows you to communicate across distances without the internet or a cell phone network. It's a hobby that combines social aspects, emergency communication, and technical learning. Let's break down the basic setup process into manageable steps:

1. **Obtain a License**

Before you can operate a ham radio, you need a license. In the U.S., the Federal Communications Commission (FCC) governs this. Start by studying for the Technician Class license, the entry-level license that gives you access to all ham radio frequencies above 30 megahertz. Various resources are available, including online courses and study guides.

2. **Choosing Your First Radio**

Types of Radios
- Handheld Transceivers (HTs): Portable and easy to use, ideal for beginners.
- Mobile Transceivers: Designed for vehicle installation but can be used at home with a proper power supply.

- Base Station Radios: Larger, more powerful units intended for home use with an external antenna.

Consider starting with an HT or a mobile transceiver for its versatility and ease of use.

Features to Look For
- Dual-band capability (2m/70cm): Allows access to two of the most popular amateur bands.
- Power output: More power can extend your communication range but will require a better antenna and power source.
- Ease of use: Look for radios with user-friendly interfaces.

3. **Setting Up the Antenna**

Your antenna is a critical component of your ham radio setup. Even the most powerful radio will perform poorly with a subpar antenna.

- For HTs: Consider upgrading from the stock rubber duck antenna to a higher quality one to improve range.
- For Mobile and Base Stations: External antennas are necessary. A simple 2m/70cm dual-band antenna is a great start for beginners. Ensure it's properly mounted and positioned for optimal performance.

4. **Power Supply**

- HTs are battery-operated, but having a spare battery or a charger is beneficial.
- Mobile radios require a 12-volt power supply. If you're using it as a base station, ensure you have a stable power source that can handle the radio's amperage requirements.

5. Learning to Use Your Radio

After setting up your station, familiarize yourself with your radio's functions and features. Practice tuning into different frequencies, listen to various broadcasts, and understand how to switch between modes (FM, AM, SSB, etc.). Most importantly, learn how to program your radio's repeaters—these are essential for extending your communication range.

6. Joining the Community

Ham radio is as much about community as it is about technology. Join local clubs or online forums to connect with experienced operators who can provide valuable insights and assistance.

7. Practicing Etiquette

Good operating practice is crucial. Listen more than you transmit, identify yourself with your call sign during communications, and be courteous to others on the air.

8. Emergency Preparedness

Many hams enjoy providing emergency communications support. Equip your station with a reliable power backup, such as a battery system or generator, and familiarize yourself with the frequencies used for emergency communications in your area.

Conclusion

Setting up your first ham radio station is the beginning of a rewarding journey into the world of amateur radio. By following these steps, you're well on your way to exploring a hobby that spans the globe and connects communities. Remember, the learning never stops in ham radio, so stay curious and keep experimenting.

Safety Considerations

Setting up your first ham radio station is an exciting step into the world of amateur radio, allowing for communication over long distances, participation in emergency networks, and exploring various frequencies. However, to ensure a safe and effective setup, several safety considerations must be adhered to.

Electrical Safety

1. Power Supply: Ensure your power supply matches the requirements of your radio equipment. Overloading electrical circuits can lead to fires or equipment damage.
2. Grounding: Proper grounding of your equipment is crucial to prevent electric shocks and protect your gear from lightning strikes. It also minimizes interference with other electronic devices.
3. Cable Management: Organize and secure cables to prevent tripping hazards and accidental disconnections. Check regularly for frayed wires or loose connections.

Antenna Safety

1. Installation: When installing antennas, especially outdoor ones, choose a location away from power lines and populated areas to minimize risk of electric shock or injury from falling equipment.

2. Height and Support: Ensure the antenna is securely mounted and the structure supporting it can withstand weather conditions. Consider the height restrictions and regulations in your area.

3. Radiation Exposure: Be mindful of RF exposure limits. Position antennas away from living spaces and use them within the recommended power levels to avoid harmful radiation exposure.

Operational Safety

1. Knowledge of Operation: Familiarize yourself with the operation of your ham radio equipment. Improper use can lead to interference with other communications, which may have legal repercussions.

2. Signal Interference: Be aware of the potential for your equipment to interfere with nearby electronic devices. Use filters and adjust your transmission power as necessary to minimize interference.

3. Emergency Protocols: Participate in ham radio emergency networks responsibly. Know the protocols for emergency communication to avoid causing confusion or congestion on emergency frequencies.

Health Considerations

1. Hearing Protection: Prolonged use of headphones at high volumes can lead to hearing damage. Adjust volume levels appropriately.
2. Physical Strain: Regular breaks can prevent strain from long periods of sitting or wearing headphones. Ergonomically arrange your station to reduce the risk of repetitive strain injuries.

Legal and Ethical Considerations

1. Licensing: Ensure you have the appropriate FCC license for your activities. Unauthorized transmission on certain frequencies can interfere with emergency services and result in penalties.
2. Respect for Band Plans: Adhere to the amateur radio band plans and respect designated uses of frequencies to promote harmony within the ham radio community.

Conclusion

By following these safety considerations, you can enjoy the world of amateur radio while minimizing risks to yourself and others. Remember, ham radio is not only about individual exploration but also about participating responsibly in a global community of amateur radio enthusiasts. Stay informed about safety practices, and don't hesitate to consult with experienced ham operators or local clubs for advice and support.

Chapter: 3 Operating Basics

Understanding Frequencies and Bands

Understanding frequencies and bands is crucial for anyone looking to operate a ham radio, as these elements form the backbone of successful communication. This comprehensive guide will help you grasp the essentials of frequencies and bands, ensuring a solid foundation for your amateur radio endeavors.

Frequencies and Their Importance in Ham Radio

Frequency refers to the number of oscillations or cycles per second of an electromagnetic wave. Measured in Hertz (Hz), frequencies dictate the behavior of radio waves and their ability to carry information over distances. In ham radio, frequencies determine not just the reach of communication but also its clarity and potential for interference.

Understanding Bands in Ham Radio

A band is a range of frequencies grouped together under a common label, such as the 2-meter band or the 20-meter band. These bands are designated for amateur radio use by regulatory bodies like the Federal Communications Commission (FCC) in the United States. Each band has unique characteristics, influenced by atmospheric conditions, time of day, and solar activity, affecting its suitability for different types of communication:

- HF Bands (3 MHz to 30 MHz): Ideal for long-distance communication, HF bands are favored for international broadcasting. Propagation relies heavily on the ionosphere, allowing signals to bounce back to Earth over great distances.
- VHF Bands (30 MHz to 300 MHz): VHF bands, including the popular 2-meter band, are excellent for local and regional communication. These frequencies typically travel in straight lines and can be enhanced with repeaters to extend their range.
- UHF Bands (300 MHz to 3 GHz): Offering even higher frequencies, UHF bands are used for short-range communication but provide the advantage of penetrating through urban environments more effectively.

Operating Basics: How to Use a Ham Radio

1. Licensing: Before operating a ham radio, obtaining a license from the FCC is mandatory. This ensures you understand the rules, operating procedures, and technical knowledge required to use the bands responsibly.

2. Choosing the Right Band: Selecting the appropriate band is crucial and depends on your communication needs. For local

chats, VHF and UHF bands are suitable, whereas HF bands are better for long-distance communication.

3. Tuning In: Using your ham radio's dial, tune into a frequency within the band you've chosen. Adjust the squelch control to minimize background noise, making it easier to identify signals.

4. Making Contact: To initiate contact, listen first to ensure you're not interrupting ongoing communication. When clear, transmit your call sign and wait for a response. Use clear and concise language, adhering to the phonetic alphabet for clarity.

5. Q Codes and Etiquette: Familiarize yourself with common Q codes (abbreviated messages) and operating etiquette. Always be respectful, patient, and willing to assist others.

6. Exploring Digital Modes: Modern ham radios support digital modes, allowing you to use computer interfaces for text, data, and image transmission. These modes can offer more efficient use of bandwidth and enhanced communication under challenging conditions.

7. Joining the Community: Ham radio is not just about technical skills but also joining a global community. Participate in clubs, contests, and emergency response activities to enhance your skills and contribute to the community.

Conclusion

Understanding frequencies and bands is just the beginning of your journey into ham radio. With knowledge, practice, and adherence to regulations and etiquette, you'll unlock the full potential of amateur radio, connecting with a diverse and supportive community worldwide. Whether for hobby, emergency communication, or exploring the airwaves, ham radio offers a rewarding experience for those willing to learn and engage.

How to Tune Your Radio

Tuning a ham radio, also known as amateur radio, can seem complex for beginners, but it's quite manageable once you grasp the basics. This guide will walk you through the fundamental steps of tuning your ham radio, ensuring you can communicate effectively and enjoy the vast world of amateur radio.

Understanding Your Ham Radio

Before tuning, familiarize yourself with your radio's components:
- VFO (Variable Frequency Oscillator): Allows you to dial into different frequencies.
- Mode Selector: Switches between different modes such as FM, AM, SSB (Single Side Band), and CW (Continuous Wave) for Morse code.
- Band Selector: Enables selection of frequency bands.
- Antenna Tuner (if available): Helps in matching the radio's output to the antenna's impedance for efficient transmission and reception.
- Squelch Control: Mutes the static noise when no signal is present.
- Volume and Gain Controls: Adjusts the loudness and sensitivity of the received signal.

Step 1: **Powering Up and Initial Setup**

1. Connect the Antenna: Ensure your antenna is properly connected to the radio. The type of antenna will depend on the bands you intend to operate on.
2. Power On: Turn on your ham radio.
3. Set the Mode: Select the appropriate mode (FM, AM, SSB, CW) based on your communication needs.
4. Volume and Squelch: Adjust the volume to a comfortable level and set the squelch to eliminate background noise.

Step 2: Selecting a Frequency

1. Choose a Band: Use the band selector to choose the frequency band you want to operate in. Bands are specific ranges of frequencies allocated for different types of communications.
2. Dialing in a Frequency: Use the VFO knob to dial into a specific frequency within the selected band. Frequencies are often chosen based on the time of day, propagation conditions, and the type of communication you intend to have.

Step 3: Tuning the Antenna (If Necessary)

1. Manual Tuner: If your setup includes a manual antenna tuner, adjust it while transmitting a low power signal until you achieve the lowest SWR (Standing Wave Ratio) reading. This indicates that the antenna is well matched to the frequency.

2. Automatic Tuner: If your radio has an automatic tuner, simply press the tune button. The radio will adjust the tuner for optimal performance on the selected frequency.

Step 4: Making Contact

1. Listening: Spend some time listening to the frequency to ensure it is clear. It's crucial not to interrupt ongoing communications.
2. Calling CQ: To initiate a call, you can say "CQ" (seek you) followed by your call sign. Speak clearly and slowly.
3. Responding: If you hear someone else calling CQ and wish to respond, wait for them to finish and then transmit your call sign during their listening period.

Step 5: QSO (Conversation)

1. Exchange Information: Typically, a QSO involves exchanging call signs, signal reports, location, and possibly other information like names or weather conditions.
2. Maintain Courtesy: Always follow the amateur radio code of conduct. Be polite, patient, and helpful to other operators.

Additional Tips

- Join a Club: Many amateur radio clubs offer mentoring to new ham operators, providing practical advice and assistance.

- Practice Makes Perfect: The more you use your radio, the more comfortable you'll become with tuning and making contacts.
- Stay Informed: Regulations and band allocations can change, so stay updated with information from the Federal Communications Commission (FCC) and the American Radio Relay League (ARRL).

By following these steps and tips, you'll be well on your way to mastering the art of tuning your ham radio, opening up a world of global communication and community. Enjoy the journey into the fascinating hobby of amateur radio!

Making Your First Contact

Making your first contact on a ham radio, often referred to as a "QSO" (Query on the air), is a thrilling milestone for any amateur radio enthusiast. This guide will walk you through the operating basics to help you make that initial leap into the airwaves smoothly and effectively.

Understanding Your Equipment

Before you attempt your first contact, familiarize yourself with your ham radio equipment. Know how to adjust the frequency, set the mode (such as FM, AM, SSB, CW), and control the volume and squelch. It's also important to understand how to use your antenna tuner to ensure your signal is as clear as possible.

Licensing and Frequencies

Ensure you have the proper licensing required to operate a ham radio in your country. In the United States, the Federal Communications Commission (FCC) regulates this. Your license class will determine which frequencies you can access. Study the band plan for your license class to know where you can legally transmit.

Preparing for Your First Contact

1. Listen First: Spend some time listening to the frequency you plan to use. This will help you get a sense of the activity and ensure you don't interrupt ongoing conversations.
2. Calling CQ: To initiate contact, you'll use the term "CQ," followed by your call sign, to indicate you're seeking a conversation. For example, "CQ CQ CQ, this is [Your Call Sign], [Your Call Sign] calling CQ and standing by."
3. Responding to Calls: If you hear someone else calling CQ and you wish to respond, wait until they finish their call and then transmit your call sign once, clearly and slowly.

Conducting the Conversation

1. Exchange of Information: Once contact is made, a typical exchange might include signal reports, location, the operator's name, and possibly a bit about the radio equipment being used. Keep it brief and to the point.
2. Using Phonetics: To ensure clarity, especially when signal conditions are poor or if there's interference, use the NATO phonetic alphabet to spell out your call sign or any critical information.
3. Signal Reports: These are usually given as two or three numbers (such as 59 or 599), indicating the signal strength and clarity. This helps the other operator understand how well their transmission is being received.

Etiquette and Best Practices

- Listen More Than You Transmit: Monitoring the bands will teach you a lot about proper operation and etiquette.
- Be Patient and Polite: Your first attempts at making contact might not go smoothly. Be patient with yourself and polite to others, regardless of any initial hiccups.
- Log Your Contacts: Keep a log of your QSOs. This is not only a requirement in some jurisdictions but also a great way to track your progress and remember the contacts you've made.

Safety and Legal Compliance

- Follow Regulations: Always operate within your license's limits and follow local and international regulations.
- Mind the Power: Use the minimum necessary power to establish and maintain contact. This is courteous to other operators and reduces the chance of interference.
- Respect Privacy: Remember that ham radio communications are public. Avoid sharing sensitive personal information.

Practice and Community Engagement

- Join a Club: Many amateur radio clubs offer mentoring for new operators, which can be invaluable as you're getting started.
- Participate in Nets and Contests: Nets (scheduled meetings on the air) and contests can be fun ways to practice operating your radio and to meet other hams.

Making your first contact is just the beginning of what can become a deeply rewarding hobby. With each QSO, you'll gain more confidence and skill. Remember, every expert operator was once a beginner, so don't hesitate to ask for help and advice from the ham radio community. Welcome to the world of amateur radio!

Q Codes and Radio Etiquette

Q Codes and radio etiquette are essential components of amateur (ham) radio communication, serving as the backbone for clear, efficient, and respectful operation. This guide outlines the fundamentals of Q Codes and radio etiquette in the context of using a ham radio, focusing on operating basics relevant to users in the United States.

Understanding Q Codes

Q Codes originated in the early 20th century as a shorthand communication method for Morse code operators. They have since been adopted by the amateur radio community for both Morse code and voice communication. Each Q Code is a three-letter code beginning with "Q," designed to convey complex information or questions quickly and efficiently.

Commonly Used Q Codes:
- QSO: Refers to a conversation between two stations.
- QRZ: Used to ask "Who is calling me?"
- QTH: Indicates the location or asks for the other station's location.
- QSL: Can either confirm receipt of transmission or request a QSL card, which is a written confirmation of a conversation between two amateur radio stations.
- QRM: Refers to interference from other signals.

- QRN: Indicates interference from natural sources, like thunder.
- QRP: Signifies low-power transmission, often asking, "Can you receive me on low power?"

Radio Etiquette

Good radio etiquette ensures that communications are carried out smoothly, without interference, and with respect for all users of the radio spectrum.

Calling CQ
- Start by checking if the frequency is in use. Listen for a while, then ask, "Is this frequency in use?" Wait for a response before proceeding.
- To initiate a call, say "CQ CQ CQ, this is [Your Call Sign], [Your Call Sign] calling CQ and standing by." Repeat as necessary but allow breaks for others to respond.

Responding to a Call
- When responding to a CQ call, clearly state the call sign of the station you're addressing, followed by "this is," then your call sign.
- Wait for them to acknowledge before proceeding with the conversation.

During the Conversation
- Keep transmissions short and to the point, allowing breaks for the other station to respond or for others to break in if necessary.

- Use plain language and standard phonetics (Alpha, Bravo, Charlie, etc.) for clarity, especially when spelling out call signs or important information.
- Respect the "listening more than transmitting" rule to avoid inadvertently interfering with ongoing transmissions.

Ending the Conversation
- When ready to end the QSO, thank the other operator for the conversation, using their call sign and yours.
- Say "73," which means "best regards" in amateur radio lingo, followed by "signing off" or "clear."

Legal and Ethical Considerations

The Federal Communications Commission (FCC) regulates amateur radio operations in the United States, and operators are expected to follow specific legal and operational guidelines. It's crucial to be familiar with these regulations, including permissible frequencies for amateur use, power limits, and identification requirements. Moreover, ethical operation—such as not interrupting ongoing conversations, not using offensive language, and offering assistance during emergencies—reflects well on the amateur radio community as a whole.

Conclusion

Understanding and adhering to Q Codes and radio etiquette are fundamental for anyone operating a ham radio. These conventions not only facilitate effective communication but also ensure that the amateur radio bands are used responsibly and respectfully. Whether you're a seasoned operator or new to ham radio, respecting these guidelines will enhance your experience and contribute positively to the global amateur radio community.

Chapter: 4 Exploring Ham Radio Activities

Contesting

Contesting in amateur radio, often referred to as "radiosport," is a competitive activity where operators attempt to make as many contacts as possible within a specific period. These events are not only fun but also sharpen the participants' operating skills, enhance their technical knowledge, and improve emergency communication capabilities. If you're interested in how to use a ham radio for contesting, here's a comprehensive guide to get you started.

Understanding Ham Radio Contesting

1. Definition and Purpose: Ham radio contesting involves operators attempting to make contact with as many stations as possible, following the specific rules of the contest. The purpose can vary from improving individual operating skills to promoting international goodwill.

2. Types of Contests: Contests can range from local and regional competitions to global ones. They may focus on a particular mode

of communication like Morse code (CW), voice (SSB), digital modes (FT8, RTTY), or a mix of several modes.

Preparing for a Contest

1. Choose the Right Equipment: A reliable transceiver capable of operating in the contest's mode (CW, SSB, digital) is crucial. Antennas should be chosen based on the bands you plan to operate on.

2. Software and Logging: Use logging software designed for contesting. It helps in logging contacts, checking for duplicates, and sometimes even assisting in operating strategy.

3. Know the Rules: Each contest has its own set of rules, including allowed frequencies, modes, power limits, and contact exchange information. Understanding these is crucial for effective participation.

4. Practice: Before diving into a contest, practice making quick, efficient contacts. Familiarize yourself with the phonetic alphabet for voice contests and your software for digital modes.

Operating in a Contest

1. Making Contacts: The key to contesting is making as many contacts as possible. Be clear and concise in your communication.

In voice contests, speak clearly; in CW and digital, ensure your setup transmits clearly.

2. Frequency Management: Find a clear frequency to operate from if calling CQ (inviting other stations to contact you), or tune the bands to make contacts with others. Be mindful of not interfering with ongoing communications.

3. Exchange Information: Most contests require the exchange of specific information, such as signal report, serial number, location, or operator's age. Ensure you understand what's required and communicate it accurately.

4. Strategy: Develop a strategy based on the contest's duration, your goals, and your station's capabilities. Some operators focus on making as many contacts as possible, while others aim for contacts in as many countries or zones as possible.

Post-Contest

1. Log Submission: After the contest, review your logs for accuracy and submit them to the contest organizers by the deadline.

2. Review and Learn: Analyze your performance to identify strengths and areas for improvement. Listening to recordings of your operation can be incredibly instructive.

3. Celebrate and Connect: Whether you win or not, participating in a contest is a significant achievement. Share your experiences with the ham radio community and connect with operators you met during the contest.

Conclusion

Ham radio contesting is a thrilling way to engage with the global amateur radio community, improve your skills, and enjoy the spirit of competition. By choosing the right equipment, understanding contest rules, and developing effective operating strategies, you can successfully participate in and enjoy ham radio contests. Remember, the ultimate goal is to have fun and learn, regardless of your score.

DXing (Long Distance Communication)

DXing, or long-distance communication, is a fascinating and rewarding aspect of amateur (ham) radio that involves making contact with other radio operators over long distances, often spanning thousands of miles. It's an activity that blends technical skills, knowledge of radio wave propagation, and operational prowess. Whether you're a seasoned ham looking to refine your DXing capabilities or a newcomer eager to embark on long-distance radio adventures, understanding the essentials of how to use a ham radio for DXing is crucial.

Understanding Ham Radio Basics

Before diving into DXing, it's important to grasp the basics of ham radio. Ham radio, or amateur radio, is a hobby and service that allows participants to communicate across distances without relying on the internet or cellular networks. Operators use various modes of communication, including voice, Morse code (CW), and digital modes, across a wide range of frequencies allocated by the FCC (Federal Communications Commission) in the United States.

Equipment and Setup for DXing

1. Transceiver: A high-quality HF (High Frequency) transceiver is crucial for DXing, capable of transmitting and receiving on the bands most favorable for long-distance communication (typically 160 to 10 meters).
2. Antenna: The choice of antenna is vital for effective DXing. Directional antennas, such as Yagi antennas, are preferred for their ability to focus signals in a particular direction, enhancing your ability to both send and receive long-distance communications.
3. Power Source: Ensure your station has a reliable power source. While most transceivers operate on standard electrical power, having a battery backup or solar options is beneficial.
4. Tuning and Filters: Familiarize yourself with your transceiver's tuning functions and filters to optimize signal reception and reduce interference.

Understanding Propagation

The ability to communicate over long distances via ham radio largely depends on understanding and utilizing radio wave propagation. Factors affecting propagation include:

- Time of Day: Depending on the frequency, some bands are better at night (e.g., 80 and 160 meters), while others perform well during the day (e.g., 10 and 20 meters).
- Solar Activity: Solar flares and sunspots can significantly affect radio wave propagation, with higher solar activity generally improving conditions.

- Atmospheric Conditions: Layers of the Earth's atmosphere, particularly the ionosphere, play a key role in the ability of radio waves to travel long distances.

Operating Practices

1. Listening: Spend more time listening than transmitting. This helps in understanding the band conditions, recognizing the presence of DX stations, and identifying the right moment to make a call.
2. Calling CQ DX: When you're ready to initiate contact, use the standard call of "CQ DX" followed by your call sign. This signals that you're looking for long-distance communication.
3. QSL Cards: Exchange QSL cards (confirmation cards) with your DX contacts. These cards serve as a physical record of your communication and are highly valued within the ham radio community.

Etiquette and Legal Considerations

- Follow Regulations: Always adhere to the frequency allocations and power limits set by the FCC or your local regulatory body.
- Respect: Practice good operating etiquette by not interrupting ongoing conversations and by using polite and respectful language.

- DX Code of Conduct: Familiarize yourself with the DX Code of Conduct, which outlines best practices for ethical and effective DXing.

Continuous Learning and Community

DXing is a continuous learning experience. Participate in ham radio clubs, online forums, and DX contests to learn from experienced operators. The ham radio community is known for its willingness to share knowledge and support fellow enthusiasts.

In conclusion, DXing is a thrilling aspect of ham radio that opens up a world of global communication. By understanding the basics of ham radio, optimizing your equipment setup, mastering the intricacies of radio wave propagation, and practicing good operating and ethical standards, you can embark on exciting long-distance communication adventures. Remember, the essence of DXing lies in the joy of discovery and connecting with people across the globe through the airwaves.

Digital Modes of Communication

Digital Modes of Communication: Exploring Ham Radio Activities

In the realm of amateur radio, or "ham" radio, the evolution of digital technology has expanded the horizons of communication. Ham radio enthusiasts, known for their innovative spirit and technical expertise, have embraced digital modes to enhance their ability to communicate worldwide. This comprehensive guide explores the various digital modes of communication within ham radio activities and offers insights into how to use a ham radio effectively in the digital age.

Understanding Digital Modes in Ham Radio

Digital modes of communication in ham radio involve the transmission of data over radio waves using digital signals. Unlike traditional analog communication, which transmits voice or Morse code, digital modes encode information into digital formats for more efficient and reliable communication. Some popular digital modes include:

- RTTY (Radio Teletype): One of the earliest digital modes, RTTY uses a simple binary coding scheme to transmit text data.

- PSK31 (Phase Shift Keying, 31 Hz Bandwidth): Highly efficient for text communication, PSK31 is favored for its ability to work well even with low power and under poor conditions.
- FT8 and FT4: These are newer, highly efficient modes designed for weak-signal communication, enabling contacts over long distances even with minimal transmission power.

How to Use a Ham Radio for Digital Communication

To venture into digital modes with your ham radio, follow these steps:

1. Choose the Right Equipment:
   - A ham radio transceiver capable of transmitting on the frequencies used for digital modes.
   - A computer with suitable software for the digital mode you wish to use (e.g., WSJT-X for FT8).
   - An interface to connect your radio to your computer, allowing them to communicate.

2. Set Up Your Station:
   - Install the digital mode software on your computer.
   - Connect the interface between your computer and your radio. This may involve audio cables for sound card modes and possibly a control cable for keying the transmitter and changing frequencies.

- Configure the software with your call sign and grid square, and adjust settings to match your radio's specifications.

3. Tune In and Decode:
   - Use the software to tune into the frequencies designated for digital modes. Each mode typically has standard frequencies on various bands.
   - Adjust your radio's settings as needed to optimize signal reception. This might include filter settings, RF gain, and other adjustments specific to your model.
   - Watch the software decode transmissions in real-time. You can see call signs, messages, and signal reports from stations around the world.

4. Make Contacts:
   - To initiate a contact (QSO), select a station from the decoded transmissions or call CQ yourself using the software.
   - Follow the protocol of the digital mode you are using. For example, FT8 has a structured exchange that includes call signs, signal reports, and acknowledgments.
   - Log your contacts. Most digital mode software can automatically log QSOs or integrate with popular logging software.

5. Explore and Experiment:
   - Experiment with different digital modes. Each has its unique characteristics and advantages.

- Participate in digital mode contests and special event stations, which are great ways to practice and increase your proficiency.
- Join online forums and local clubs to share experiences and learn from other ham radio operators.

Final Thoughts

Digital modes of communication in ham radio open up a new world of possibilities for amateur radio enthusiasts. By combining the traditional aspects of ham radio with modern digital technology, operators can enjoy more efficient, reliable, and exciting ways to connect across the globe. Whether you're a seasoned ham looking to expand your capabilities or a newcomer eager to dive into digital modes, the ham radio community welcomes you with endless opportunities for learning and exploration.

Emergency Communications and Public Service

Emergency Communications and Public Service: Exploring Ham Radio Activities

In the realm of emergency preparedness and public service, Ham (Amateur) Radio stands out as an indispensable tool, providing reliable communication when all else fails. As natural disasters and unforeseen events disrupt conventional communication channels, Ham Radio operators play a crucial role in emergency response and community service. This comprehensive guide delves into the basics of Ham Radio, its significance in emergency communications, and provides a step-by-step approach to using these radios effectively.

Understanding Ham Radio

Ham Radio, or Amateur Radio, is a form of communication that uses radio frequencies to exchange messages, conduct experiments, and enhance technical skills. Governed by the Federal Communications Commission (FCC) in the United States, Ham Radio is not just a hobby but a means to serve the community, especially in times of emergency.

The Significance of Ham Radio in Emergency Communications

1. Reliability: When cell phones, internet, and other communication infrastructures fail, Ham Radios continue to operate, making them invaluable during disasters.
2. Versatility: Ham Radio can transmit voice, data, and Morse code across the globe without relying on external networks.
3. Community Service: Ham Radio operators often participate in public service events, such as marathons and festivals, to provide communication support.

Getting Started with Ham Radio

To operate a Ham Radio, one must obtain a license from the FCC. The process involves studying for and passing an examination that tests knowledge on radio theory, regulations, and operating practices.

How to Use a Ham Radio: A Step-by-Step Guide

Step 1: Setting Up Your Equipment
- Choose a Radio: Start with a simple handheld transceiver (HT) for local communication. For broader reach, consider base station or mobile units.
- Antenna Setup: The right antenna is crucial for effective communication. Higher antennas generally provide better coverage.

- Power Source: Ensure your radio is charged or connected to a reliable power source.

Step 2: Understanding Frequencies and Channels
- Familiarize yourself with the frequencies allocated for Amateur Radio use. Each band has its characteristics and is suitable for different types of communication.

Step 3: Making Your First Call
- Listen before you transmit to ensure the frequency is clear.
- Press the transmit (PTT) button and introduce yourself using your call sign, followed by the standard phrase "listening" or "CQ" to seek communication.

Step 4: Joining Emergency Networks
- Many communities have Amateur Radio Emergency Services (ARES) or Radio Amateur Civil Emergency Service (RACES) networks. Joining these can provide structured opportunities to support emergency communications.

Step 5: Practice and Participation
- Regular participation in public service events and emergency drills will enhance your skills.
- Joining a local Ham Radio club can provide mentorship and opportunities to practice.

Conclusion

Ham Radio is not only a hobby but a lifeline during emergencies, providing an essential communication channel that can operate independently of traditional infrastructure. By understanding the basics of Ham Radio, obtaining a license, and practicing regularly, individuals can contribute significantly to their communities' safety and well-being. Whether it's facilitating communication during a disaster or offering support at a local event, Ham Radio operators are invaluable assets to public service and emergency response efforts.

Chapter: 5 Antennas and Propagation

Types of Antennas

When using a ham radio, the choice of antenna can significantly impact both the quality and reach of your communications. Antennas are the interface between the radio waves propagating through space and the electric signals used in radio equipment. Understanding the types of antennas and how they relate to ham radio use can enhance your communication experience. Here's a detailed exploration of various antenna types in the context of ham radio usage:

1. **Dipole Antennas**

- Description: The dipole antenna is the most basic and commonly used antenna in ham radio. It consists of two metal rods or wires of equal length, oriented end to end with a small gap in the middle where it's connected to the transmission line.
- Ham Radio Use: Ideal for beginners, it's typically used for HF bands. Its simplicity makes it easy to construct and adjust for different frequencies. Dipole antennas can be installed horizontally, vertically, or in an inverted-V configuration, affecting their radiation pattern and impedance.

2. **Yagi-Uda Antennas**

- Description: A directional antenna consisting of a dipole (driven element) with additional passive elements: reflectors and directors, arranged on a beam. The number of elements can vary, affecting the antenna's gain and directivity.
- Ham Radio Use: Excellent for VHF and UHF bands, Yagi antennas are favored for chasing distant contacts (DXing) and satellite communication. Their directional nature allows for focused transmission and reception, enhancing signal strength in a particular direction.

3. **Vertical Antennas**

- Description: These antennas stand upright and radiate signals 360 degrees around the antenna, making them omnidirectional. They vary in complexity from simple quarter-wave monopoles to sophisticated phased arrays.
- Ham Radio Use: Vertical antennas are useful for local and medium-distance contacts on lower frequency bands. They're particularly beneficial in areas with limited space and for mobile operations, including maritime and mobile stations.

4. **Loop Antennas**

- Description: Loop antennas can range from small, circular designs to large horizontal loops encompassing hundreds of feet of wire. Their performance is characterized by a low noise floor and a distinctive radiation pattern.
- Ham Radio Use: They're versatile, working well for both receiving and transmitting across a wide range of frequencies. Small magnetic loop antennas are especially popular for amateur radio operators with limited space.

5. Quad Antennas

- Description: Similar to the Yagi, with elements made of loops rather than straight rods, forming a square or rectangular shape. They can be either monoband or multiband and are known for their high gain.
- Ham Radio Use: Quads are often used by enthusiasts looking to make long-distance contacts due to their excellent gain and directivity properties. They require more space than Yagi antennas but can offer superior performance in certain conditions.

6. Beam Antennas

- Description: Beam antennas are a category that includes both Yagi and Quad antennas. The term "beam" refers to the antenna's ability to concentrate signals in a specific direction, offering high directivity and gain.

- Ham Radio Use: Beam antennas are the go-to choice for serious DXers and contesters looking to maximize their signal strength and directionality.

7. **Wire Antennas**

- Description: This broad category includes various designs like dipoles, inverted V, and long wires, made from lengths of wire cut to specific frequencies.
- Ham Radio Use: Wire antennas are favored for their simplicity, cost-effectiveness, and versatility. They can be easily erected at home, in the field, or for portable operations, making them suitable for a wide range of ham radio activities.

8. **Magnetic Loop Antennas**

- Description: A small, highly efficient antenna that consists of a loop of wire or tubing through which a high current flows. It's known for its small size and narrow bandwidth.
- Ham Radio Use: Ideal for apartment dwellers or those with limited space, magnetic loop antennas can be used for both receiving and transmitting on HF bands, though they typically require tuning for each frequency used.

Conclusion

Choosing the right antenna for your ham radio setup involves considering your available space, the bands you wish to operate on, and whether you're more interested in local or long-distance (DX) communications. Each antenna type has its advantages and applications, and many ham radio operators enjoy experimenting with different antennas to find the perfect match for their interests and conditions.

Basic Antenna Theory

"Basic Antenna Theory" in relation to ham radio operation is a fundamental topic that intertwines the physical principles of antennas with the practical aspects of amateur radio communication. Understanding antenna theory can significantly enhance the efficiency, reach, and overall satisfaction of using ham radios. This comprehensive guide aims to bridge the gap between theoretical knowledge and practical application for ham radio enthusiasts.

Understanding Antennas

At its core, an antenna is a device designed to transmit or receive electromagnetic waves. Antennas convert electrical power into radio waves when transmitting and vice versa when receiving. The effectiveness of this conversion process significantly affects the communication range and clarity.

Types of Antennas

1. Dipole Antennas: Often the starting point for most ham operators, a dipole is a simple, effective antenna that's easy to construct. It consists of two pieces of wire or metal rods, with the feed line connected in the middle.

2. Yagi-Uda Antennas: Known for their directional capabilities, Yagi antennas are ideal for communicating over long distances in a specific direction. They consist of a driven element (dipole), a reflector, and one or more directors.

3. Vertical Antennas: These antennas are ground-dependent and are known for their omnidirectional patterns, making them suitable for various communication needs. They're particularly effective for HF bands.

4. Loop Antennas: Loops can be small or large and are known for their efficiency and directionality. They're particularly useful for receiving, offering a quieter alternative to other designs.

5. Beam Antennas: For operators looking to maximize directional gain, beam antennas (including Yagi designs) are the go-to choice, especially for contests and DXing (long-distance communication).

Antenna Properties

1. Polarization: Refers to the orientation of the electromagnetic wave's electric field. In ham radio, antennas can be vertically or horizontally polarized, affecting how well they receive signals from other antennas based on matching polarization.

2. Bandwidth: This is the range of frequencies over which the antenna can effectively operate. A wider bandwidth means the antenna can handle a broader range of frequencies, which is useful for bands with wide allocations.

3. Gain: A measure of an antenna's ability to direct or concentrate radio frequency energy in a particular direction or pattern. Higher gain antennas can send and receive signals over greater distances but are often more directional.

4. Impedance: The resistance an antenna presents to the flow of RF energy. Most ham radios are designed to work with an antenna impedance of 50 ohms. Matching the impedance of the antenna to the transmitter is crucial for efficient power transfer.

Setting Up and Using Antennas in Ham Radio

1. Choosing the Right Antenna: Consider your communication goals (local vs. long-distance), available space, and the frequencies you plan to operate on. Often, a simple dipole is a great starting point.

2. Antenna Placement and Orientation: Higher is usually better for line-of-sight VHF/UHF communications, while HF antennas might have different height requirements based on the desired angle of radiation for DX or NVIS (Near Vertical Incidence Skywave) communications.

3. Tuning the Antenna: Many antennas require tuning to operate efficiently at the desired frequencies. This can involve physical adjustments or using an antenna tuner to match the antenna's impedance to the radio.

4. Safety Considerations: Always ensure antennas are well clear of power lines, and take care when installing antennas at height. RF exposure limits should be observed to keep you and others safe.

Practical Tips

- Experiment: Antenna building and experimentation are core aspects of the ham radio hobby. Don't hesitate to try different antennas and configurations.
- Join a Club: Many ham radio clubs offer workshops and mentorship for building and optimizing antennas.
- Resources and Learning: Books, online forums, and resources from the Amateur Radio Relay League (ARRL) can provide invaluable guidance and technical information.

In conclusion, mastering basic antenna theory and its practical applications can significantly enhance your ham radio experience. Whether you're communicating across town or around the world, the right antenna setup is key to successful and enjoyable amateur radio operations.

Understanding Radio Wave Propagation

Understanding radio wave propagation is crucial for effectively using a ham radio, as it directly impacts how, when, and where we can communicate over amateur radio frequencies. This comprehensive guide explores the essentials of radio wave propagation, the role of antennas, and practical insights for ham radio enthusiasts.

What is Radio Wave Propagation?

Radio wave propagation refers to how radio waves travel from the transmitter to the receiver. This journey can be affected by various factors, including the frequency of the wave, the atmosphere, and physical obstacles. Understanding these factors helps ham radio operators choose the right equipment and settings for their communication needs.

Key Factors Influencing Propagation

1. Frequency: The frequency of a radio wave affects how it travels. Lower frequencies (like those on the HF band) can travel longer distances by reflecting off the ionosphere, while higher frequencies (VHF and above) generally travel in straight lines and are suited for line-of-sight communication.

2. Atmosphere: The Earth's atmosphere, especially the ionosphere, plays a significant role in radio wave propagation. Layers of the ionosphere can reflect radio waves back to Earth, allowing them to cover distances far beyond the horizon.
3. Obstacles: Physical obstacles such as buildings, trees, and terrain features can block or reflect radio waves, impacting their ability to reach the receiver.

Antennas and Their Importance

Antennas are critical components of ham radio setups. They convert electrical signals into radio waves and vice versa. The choice of antenna can significantly affect the quality and reach of communication. There are several types of antennas, each suited for different bands and propagation conditions:

- Dipole Antennas: Simple yet effective, ideal for beginners and versatile across various bands.
- Yagi-Uda Antennas: Highly directional, offering excellent gain for distant contacts, especially on the VHF and UHF bands.
- Vertical Antennas: Good for long-distance communication on the HF bands, as they emit low-angle radiation favorable for ionospheric reflection.

Maximizing Ham Radio Use Through Propagation Understanding

1. Time of Day: Propagation conditions vary throughout the day. For instance, the best time for HF band operation is during the night or early morning when the ionosphere's reflective properties are enhanced.
2. Seasonal Variations: Seasonal changes affect propagation, with some bands performing better in summer and others in winter, primarily due to changes in the ionosphere.
3. Solar Activity: Solar flares and sunspots can significantly affect propagation by either enhancing or degrading the ionosphere's ability to reflect radio waves.

Practical Tips for Ham Radio Operators

- Monitor Propagation Forecasts: Use online resources and tools to stay informed about current propagation conditions and solar activity.
- Experiment with Antennas: Try different antennas and orientations to see what works best for your desired communication range and conditions.
- Join the Community: Engage with other ham radio operators. They can offer valuable insights and firsthand experiences on maximizing propagation advantages.

Conclusion

Understanding radio wave propagation and its interplay with antennas is fundamental for every ham radio operator. By mastering these concepts, you can significantly enhance your ability to communicate over various distances, conditions, and frequencies. Remember, successful ham radio operation is as much about knowledge as it is about experimentation and adaptation to the ever-changing propagation conditions.

Chapter: 6 Advanced Operation Techniques

Adjusting Your Radio for Optimal Performance

Adjusting your ham radio for optimal performance involves fine-tuning various settings and understanding advanced operational techniques. Ham radio, or amateur radio, offers a unique and powerful way of communicating across distances without relying on the internet or cellular networks. Here's a comprehensive guide on how to adjust your radio for optimal use.

Understanding Your Equipment

Before adjusting your ham radio, it's crucial to familiarize yourself with its components and how they work. Know the specifics of your model, including its transmitter, receiver, and antenna system. Consult the manual for basic operations and specifications.

Basic Settings Adjustments

1. Frequency Selection: Choose your frequency based on the time of day, band conditions, and intended distance of

communication. Lower frequencies (160m, 80m, 40m bands) are better at night, while higher frequencies (20m, 15m, 10m bands) perform well during the day.

2. Mode Selection: Decide between modes like FM, AM, SSB, or digital modes (PSK31, FT8). Each has its advantages depending on the communication need and band conditions.

3. Filter Settings: Adjust your radio's filters to reduce noise and interference. Narrow the filter bandwidth for CW (Morse code) and digital modes, and widen it for voice communication.

Advanced Operational Techniques

1. Antenna Tuning: Use an antenna tuner to match the impedance of your antenna to your radio. This maximizes power transfer and efficiency. Understand the difference between manual and automatic tuners and how to operate them effectively.

2. SWR (Standing Wave Ratio) Measurement: Regularly check the SWR to ensure your antenna system is properly tuned. A low SWR means more power is being transmitted, and less is reflected back into the radio, reducing the risk of damage.

3. Signal Processing: Utilize your radio's built-in DSP (Digital Signal Processing) to filter out noise and enhance signal quality.

Adjust the DSP settings to improve the clarity of received signals and reduce fatigue during long operation periods.

4. Gain Control: Master the use of RF gain and AF gain controls. RF gain adjusts the sensitivity of the receiver to weak signals, while AF gain controls the volume of the audio output. Balancing these can greatly improve the signal-to-noise ratio.

5. Use of Amplifiers: For long-distance communication, especially during poor band conditions, an amplifier can boost your signal's strength. Ensure your amplifier is compatible with your radio and understand the legal limits on power output.

6. Operating Split: In crowded band conditions, operating "split" allows you to listen on one frequency while transmitting on another. This technique is especially useful in DX (long-distance) contacts and pile-ups.

Maintenance and Safety

- Regularly inspect your antenna, cables, and connectors for wear and tear.
- Ensure proper grounding of your equipment to avoid electrical hazards.
- Be mindful of RF exposure and adhere to safety guidelines to protect yourself and others.

Continuous Learning

- Join ham radio clubs and communities for advice and mentorship.
- Participate in ham radio contests to sharpen your operating skills.
- Stay updated on the latest advancements in ham radio technology and regulations.

Adjusting your ham radio for optimal performance is a blend of technical know-how, practical experience, and continuous learning. By mastering these settings and techniques, you'll enhance your communication capabilities, making your ham radio operation both more efficient and enjoyable.

Using Amplifiers for Increased Power

Using amplifiers with ham radios is a technique that can significantly enhance the power and reach of your transmissions. This approach, however, demands careful consideration of several technical aspects to ensure legal compliance and optimal performance. Here's a comprehensive guide to using amplifiers for increased power in ham radio operations:

1. **Understanding Amplifiers**

Amplifiers, or "linear amplifiers," boost the power of the signal generated by your ham radio. They're particularly useful in scenarios where you need extended range or when operating in challenging conditions. The key specifications to consider include gain (the amount of signal amplification), input and output impedance, and the amplifier's operating frequency range.

2. **Legal Considerations**

Before incorporating an amplifier, it's crucial to understand the legal limitations set by the Federal Communications Commission (FCC) in the United States. The FCC regulates the maximum power output levels to prevent interference with other communications. Always verify the legal power limits for your license class and ensure your setup complies.

3. **Choosing the Right Amplifier**

Select an amplifier that matches your ham radio's capabilities and your operational needs. Consider the following factors:
- Compatibility: Ensure the amplifier supports the modes of operation you plan to use (e.g., CW, SSB, digital modes).
- Power Requirements: The amplifier should match the output power your radio can safely provide as input.
- Frequency Range: The amplifier must cover the frequency bands you intend to operate on.

4. **Integration with Your Setup**

Proper integration of an amplifier into your ham radio setup involves physical connections and ensuring the radio and amplifier communicate effectively. This usually means connecting the radio's output to the amplifier's input and linking any control or ALC (Automatic Level Control) circuits to prevent signal distortion.

5. **Operating Techniques**

- Tuning and Load Matching: To maximize efficiency and minimize SWR (Standing Wave Ratio), tune your amplifier and antenna system for optimal impedance matching.
- Managing Heat: Amplifiers generate heat, so adequate cooling is necessary to prevent damage. Use fans or heat sinks as required.

- Drive Level Adjustment: Adjust the drive level (the signal input into the amplifier) to avoid overdriving and potentially damaging the amplifier or creating interference.

6. Safety Precautions

Working with amplifiers involves high voltages and currents. Always prioritize safety by:
- Turning off and unplugging equipment before making connections.
- Using proper grounding techniques to protect against electrical hazards.
- Being cautious of RF exposure, especially with higher power levels.

7. Maintenance and Troubleshooting

Regular maintenance ensures your amplifier's longevity and reliability. Keep connections clean and secure, and periodically check for signs of wear or damage. Familiarize yourself with common issues like overheating, erratic SWR readings, or unexpected power drops, and learn how to troubleshoot these problems.

8. Ethical Operation

Always operate your amplified ham radio setup with consideration for other users of the radio spectrum. Avoid unnecessary transmissions on busy channels, and strive to use the minimum power necessary to establish and maintain communication.

Conclusion

Incorporating an amplifier into your ham radio setup can significantly enhance your ability to communicate over long distances and in challenging conditions. However, this comes with the responsibility of understanding and adhering to legal requirements, technical specifications, and operational best practices. With the right knowledge and approach, using an amplifier can be a rewarding aspect of advanced ham radio operation, opening up new possibilities for communication and exploration of the radio spectrum.

Advanced Digital and Satellite Communications

Advanced digital and satellite communications have significantly expanded the capabilities and range of ham (amateur) radio operations, providing enthusiasts with more ways to connect, experiment, and provide emergency communications. When incorporating advanced operation techniques into ham radio usage, one delves into sophisticated areas like digital modes, satellite communication, and even linking to the Internet. Here's a detailed exploration of how to use a ham radio within the context of advanced digital and satellite communications.

Understanding the Basics

Before diving into advanced operations, it's crucial to have a solid understanding of ham radio basics, including obtaining the proper license, understanding radio frequency principles, and being proficient with basic radio equipment.

Digital Modes of Communication

Digital modes have transformed ham radio, allowing for more efficient use of bandwidth and providing clearer communications over long distances without the need for high power. Modes like PSK31, FT8, and JT65 enable text and data transmission over the airwaves, often with software assistance.

How to Use Digital Modes:
1. Equipment Setup: Ensure your radio supports the desired digital mode. You'll likely need a computer with sound card interface connected to your radio.
2. Software: Install digital mode software like WSJT-X for modes like FT8 or JT65, or fldigi for PSK31 and other digital modes.
3. Tuning and Operations: Use the software to tune your radio to the digital mode frequencies. Follow on-screen prompts to transmit and receive messages.

Satellite Communications

Amateur radio satellites (AMSAT) offer a unique way to communicate across the globe by bouncing signals off satellites in orbit.

How to Communicate via Satellite:
1. Understanding Orbits: Learn about the satellite's orbit to know when it will be overhead and accessible from your location.
2. Equipment: A dual-band VHF/UHF radio is essential. You may also need a directional antenna to track the satellite across the sky.
3. Making Contacts: Tune your radio to the satellite's uplink frequency to send your signal and listen on the downlink frequency. Adjust for Doppler shift as the satellite moves.

Internet Linking Technologies

Technologies like EchoLink and D-STAR allow ham operators to link their radios to the Internet, extending the reach of their communications globally.

Using Internet Linking:
1. Setup: For EchoLink, you need a computer with Internet, software from the EchoLink website, and verification of your ham radio license.
2. Operation: Once set up, you can connect to repeaters or other users worldwide through your computer or a compatible radio.

Advanced Operation Techniques

1. DXing: This involves communicating with distant stations. It requires understanding of propagation, optimal times for communication, and often a good antenna setup.
2. Contesting: Participating in contests to make as many contacts as possible within a specific time frame, testing your operational skills and station setup.

Best Practices

- Continuous Learning: The world of amateur radio is ever-evolving, with new technologies and modes being developed. Stay informed through clubs, forums, and publications.

- Safety and Etiquette: Always adhere to the amateur radio code of conduct and operate your station safely, especially when dealing with high power transmissions and antenna installations.
- Community Engagement: Ham radio is not just about the technology but also about the community. Engage with local clubs and online communities to share knowledge and experiences.

Advanced digital and satellite communications in ham radio offer a blend of traditional radio principles with cutting-edge technology, providing a rich platform for communication, experimentation, and service. Whether you're communicating through digital modes, bouncing signals off satellites, or linking your radio to the Internet, the world of amateur radio has something for everyone willing to explore its depths.

Chapter: 7 Joining the Ham Radio Community

Finding and Joining Local Clubs

Joining the Ham Radio Community: A Guide to Finding and Joining Local Clubs

Ham radio, also known as amateur radio, is a popular hobby and service that brings people together through communication over radio waves. Whether you're an experienced operator or new to the world of amateur radio, joining a local club can significantly enhance your experience. Here's a comprehensive guide to finding and joining local Ham radio clubs, including tips on how to use a Ham radio.

Step 1: Understanding Ham Radio

Before diving into the local club scene, it's crucial to understand what Ham radio is and what it entails. Ham radio involves using various types of radio equipment to communicate with other amateur radio enthusiasts across the globe. It requires obtaining a license from the Federal Communications Commission (FCC) in the United States, which involves passing an examination that tests your knowledge of electronics, radio theory, and FCC regulations.

Step 2: Finding Local Clubs

Local Ham radio clubs are the backbone of the amateur radio community. They provide support, resources, and a sense of camaraderie for enthusiasts. Here's how to find them:

- ARRL Club Search: The American Radio Relay League (ARRL) offers a club search tool on their website. This tool allows you to find clubs by state, city, or ZIP code.
- Social Media and Online Forums: Platforms like Facebook, Reddit, and specific amateur radio forums can be great places to find local clubs. Look for groups or threads dedicated to Ham radio in your area.
- Hamfests and Conventions: These events are gatherings of amateur radio enthusiasts and are often sponsored by local clubs. They can provide a direct route to finding a club and meeting members in person.
- Word of Mouth: Ask around. If you know someone who is involved in amateur radio, they can likely point you towards a local club.

Step 3: Joining a Club

Once you've found a club that interests you, the next step is to join. Here's what you can typically expect:

- Attend Meetings: Most clubs have regular meetings. Attend a few to get a feel for the club's activities and members.
- Membership Application: If you decide the club is a good fit, you'll need to fill out a membership application. Some clubs may require a small fee to join.
- Participate: Clubs offer various activities, from educational programs and workshops to public service events. Participation is key to getting the most out of your membership.

Step 4: Learning to Use a Ham Radio

After joining a club, you'll have access to a wealth of knowledge on how to use Ham radio effectively. Here are some basics to get you started:

- Get Licensed: If you haven't already, obtain your Ham radio license from the FCC. Your club can provide resources and study guides to help you prepare for the exam.
- Equipment Basics: Learn about the different types of radio equipment, how to set up your station, and the basics of operating it. Club members can offer advice and hands-on demonstrations.
- On-Air Etiquette: Understanding the protocols and etiquette of communicating on the airwaves is crucial. Experienced club

members can guide you through your first contacts and help you become a proficient operator.
- Explore Different Modes: Ham radio isn't just about talking. You can also use digital modes, Morse code (CW), and even bounce signals off satellites. Explore different aspects of the hobby with the help of your club.

Conclusion

Joining a local Ham radio club can transform your amateur radio experience. It offers the opportunity to learn, grow, and share in a community of like-minded individuals. From finding the right club to making your first contact on the air, the journey into the world of Ham radio is one of continuous learning and discovery. With the support of a local club, you'll be well on your way to mastering the airwaves.

Participating in Nets and Roundtables

Participating in nets and roundtables is a significant aspect of joining the ham radio community, offering both new and experienced operators valuable opportunities for learning, camaraderie, and emergency communication preparedness. This guide provides a comprehensive overview of how to engage effectively in these activities, enhancing your ham radio experience.

Understanding Nets and Roundtables

Nets, or network sessions, are scheduled meetings of ham radio operators over the air. They serve various purposes, including emergency preparedness, technical discussions, or simply social interaction. Nets can be local, regional, or even international, depending on the frequencies and modes used.

Roundtables are more informal gatherings, often without a strict schedule or script, where operators exchange information, experiences, and stories. They foster a sense of community and provide a platform for mentorship and learning.

Getting Started

1. **Licensing and Equipment**

Before participating, ensure you have the appropriate ham radio license and a properly set-up station. Familiarize yourself with your equipment's operation, focusing on the frequencies and modes most commonly used for nets and roundtables in your area.

2. **Finding Nets and Roundtables**
- Local Clubs: Joining a local ham radio club can provide you with a schedule of nets and roundtables.
- Online Resources: Websites and online forums offer listings and schedules for various nets and roundtables.
- Listen: Simply tuning into frequencies known for nets and roundtables can help you find sessions that match your interests.

Participating in Nets

1. **Check-in Process**
Most nets have a specific protocol for check-ins. Listen first to understand the net control station's (NCS) instructions. When it's your turn, typically, you'll need to provide your call sign, name, and location.

2. **Net Etiquette**
- Listen More Than You Talk: Especially when new, listening can help you understand the net's flow and etiquette.
- Follow the NCS Instructions: The NCS coordinates the net, so following their guidance is crucial for smooth operation.

- Keep Transmissions Short: To allow time for others, keep your comments concise.

Engaging in Roundtables

1. **Joining In**

Unlike nets, roundtables may not have a formal check-in process. When you find a roundtable, listen for a while to get a sense of the conversation's topic and tone. When there's a natural pause, introduce yourself with your call sign and name.

2. **Conversation Flow**

In roundtables, the conversation can be more free-flowing. While it's important to be respectful and not monopolize the conversation, feel free to share your experiences and ask questions.

Tips for Effective Communication

- Clear Speech: Speak clearly and at a moderate pace to ensure you're understood, especially when conditions might be less than ideal.
- Use Standard Phrasing: In nets, especially, using standard phrasing and protocols helps maintain clarity and efficiency.
- Be Patient and Respectful: Not every transmission will go perfectly. Patience and respect go a long way in building positive relationships within the ham radio community.

Conclusion

Participating in nets and roundtables is a rewarding aspect of being part of the ham radio community. It offers opportunities for learning, emergency communication preparedness, and making lifelong friendships. By understanding the basics of how to participate effectively and respectfully, you'll enhance your experience and contribute positively to the community.

Ham Radio Conferences and Events

Joining the ham radio community involves more than just understanding the technical aspects of operating ham radios; it's also about connecting with other enthusiasts, sharing knowledge, and keeping up with the latest trends and regulations in the field. Ham radio conferences and events play a crucial role in this, offering both newcomers and seasoned operators valuable opportunities to learn, network, and contribute to the broader community. Here's a comprehensive guide on how these events can enhance your journey into ham radio, particularly focusing on how to use a ham radio effectively.

Understanding Ham Radio Conferences and Events

Types of Events

1. Local Club Meetings: These are great starting points for beginners. Local ham radio clubs host regular meetings, workshops, and "Elmer" sessions where experienced hams mentor newcomers on various topics, including how to use ham radios.

2. Hamfests: Part swap meet, part social gathering, hamfests are where enthusiasts buy equipment, attend workshops, and meet other hams. They often include demonstrations on how to set up and operate different kinds of ham radios.

3. Conventions: Larger than hamfests, these events can span several days and include extensive programs with workshops, keynote speakers, and exhibitions showcasing the latest technology in ham radio.

4. Special Event Stations: These are temporary stations set up during significant local or national events, offering a unique opportunity to practice operating radios under different conditions and to engage with the broader public.

5. Contests and DXpeditions: Though not conferences, these competitive events challenge operators to make as many contacts as possible under specific conditions, sharpening technical skills and understanding of radio operation.

How These Events Enhance Ham Radio Usage

1. Educational Workshops and Seminars: Learn from experienced operators about antenna design, propagation, digital modes, and emergency communications. Direct interaction allows for immediate feedback and clarification of complex concepts.

2. Networking with Experienced Hams: Building relationships with experienced operators can provide mentorship opportunities and the chance to learn hands-on tips and tricks not found in manuals or guides.

3. Equipment Demonstrations: Seeing and sometimes operating state-of-the-art equipment can enhance your understanding of the possibilities within ham radio and inspire investments in your setup.

4. Q&A Sessions: These sessions can demystify aspects of radio operation that might be confusing for beginners, from navigating regulatory requirements to optimizing radio settings for various conditions.

5. Hands-On Learning: Many events offer hands-on workshops where you can practice setting up equipment, troubleshooting, and making contacts under the guidance of experienced operators.

Getting Involved

To start, identify local clubs and national organizations such as the American Radio Relay League (ARRL) in the U.S. These bodies often list upcoming events on their websites and provide resources for beginners. Subscribing to ham radio publications and joining online forums can also keep you informed about events and developments in the field.

Conclusion

Ham radio conferences and events offer invaluable resources for anyone looking to join the ham radio community. They provide a platform for learning, sharing, and connecting with other enthusiasts, which is essential for mastering the art of ham radio operation. Whether you're a beginner eager to learn the basics or an experienced operator looking to expand your knowledge and skills, participating in these events can significantly enrich your ham radio journey.

Chapter: 8 Maintenance and Troubleshooting

Routine Maintenance Checks

Routine maintenance checks are essential for ensuring the optimal performance and longevity of your ham radio equipment. Regular maintenance and troubleshooting can prevent minor issues from becoming major problems, keeping your communication clear and effective. Here's a comprehensive guide on routine maintenance checks for ham radio operators:

1. **Visual Inspection**

- Exterior Cleaning: Regularly dust and clean the exterior of your radio with a dry, soft cloth. Avoid using harsh chemicals or abrasives.
- Connection Check: Inspect all external connections for wear or damage. Ensure that plugs and jacks are secure and free from corrosion.

2. **Internal Inspection**

- Ventilation: Ensure that the radio's ventilation ports are clear of dust and debris to prevent overheating.

- Circuit Boards and Wiring: Look for signs of wear, loose connections, or corrosion on circuit boards and wiring. Use a can of compressed air to gently remove dust.

3. Power Supply and Battery Maintenance

- Power Supply Inspection: Check power supply cords and connections for damage or wear. Ensure that the power supply is providing the correct voltage to your radio.
- Battery Health: For portable radios, regularly check the battery health. Replace batteries that no longer hold a charge or show signs of swelling.

4. Antenna System

- Antenna Inspection: Regularly inspect your antenna for physical damage or wear. Ensure that all connections are tight and free of corrosion.
- SWR Check: Use an SWR meter to check the standing wave ratio. High SWR readings can indicate issues with the antenna, cable, or connectors.

5. Software Updates

- Firmware Updates: Check the manufacturer's website for firmware updates. Keeping your radio's firmware up to date can improve performance and add new features.

6. Performance Testing

- Audio Quality: Transmit and receive test messages to check for clarity. Listen for any distortion or interference.
- Sensitivity and Selectivity: Use known weak signals to test the receiver's sensitivity. Check that your radio can distinguish between closely spaced signals.

7. Troubleshooting Common Issues

- Poor Transmission Quality: Check antenna connections, inspect the microphone and wiring, and ensure the radio is correctly tuned.
- Receiver Issues: If the receiver is underperforming, check the antenna system, ensure the radio is not being overloaded by strong nearby signals, and adjust the RF gain control as necessary.
- Interference: Use filters or adjust the antenna placement to minimize interference. Identify the source of interference if possible and take appropriate measures to mitigate it.

8. Documentation and Logs

- Maintenance Logs: Keep detailed logs of all maintenance activities, including dates, actions taken, and any parts replaced. This history can be invaluable for troubleshooting future issues.

- Operating Manual: Familiarize yourself with the operating manual for your specific model. The manual can provide guidance on routine maintenance and troubleshooting.

Conclusion

Routine maintenance checks are a crucial aspect of ham radio operation. By regularly inspecting and maintaining your equipment, you can ensure reliable communication and extend the life of your radio. Remember, prevention is always better than cure, so dedicate time to perform these checks regularly and address any issues promptly.

Common Issues and How to Resolve Them

Ham radio, or amateur radio, is a popular hobby and service that allows people to communicate across distances without relying on the internet or cellular networks. Like any technology, ham radio can have its share of issues, but many can be resolved with a bit of knowledge and troubleshooting. Here are some common problems and their solutions:

1. **Poor Reception**
- Cause: Poor reception can be caused by antenna issues, interference from electronic devices, or environmental factors.
- Resolution: Ensure your antenna is properly installed and positioned. Consider upgrading to a higher quality antenna or using a directional antenna to improve reception. Turn off electronic devices that may cause interference.

2. **High Standing Wave Ratio (SWR)**
- Cause: A high SWR indicates a mismatch between the antenna and the transmitter, which can lead to poor performance and even damage the radio.
- Resolution: Use an SWR meter to check the ratio. Adjust the length of the antenna, replace the coaxial cable if damaged, or use an antenna tuner to improve the match.

3. **Transmitter Not Keying**

- Cause: This issue can occur due to problems with the microphone, the keying circuit, or the power supply.
- Resolution: Check the microphone and its connection. Ensure the power supply is adequately providing power to the transmitter. Inspect the keying circuit for any faults.

4. Interference From Other Stations
- Cause: Interference can happen if stations are operating too closely in frequency or if there's excessive signal bleed-over.
- Resolution: Try changing frequencies to avoid crowded channels. Use filters to minimize interference. Adjust the squelch setting on your radio to help suppress unwanted signals.

5. Unable to Connect to Repeaters
- Cause: This might be due to incorrect repeater settings, such as the offset frequency or sub-audible tones not being correctly programmed.
- Resolution: Double-check the repeater's settings, including the offset and tone frequencies. Make sure your radio is programmed according to the repeater's specifications.

6. Audio Quality Issues
- Cause: Poor audio can be caused by incorrect microphone use, problems with the radio's audio settings, or issues with the antenna system.
- Resolution: Speak clearly and directly into the microphone at an appropriate volume. Adjust the radio's audio settings, such as the

microphone gain, if available. Check the antenna system for any issues that might affect transmission quality.

7. Battery Drain
- Cause: Rapid battery depletion can occur due to leaving the radio on for extended periods, using high power modes, or having an old or damaged battery.
- Resolution: Turn off the radio when not in use. Use lower power modes whenever possible. Replace old or damaged batteries with new ones.

8. Difficulty Programming the Radio
- Cause: Programming can be complex due to the multitude of settings and options available on ham radios.
- Resolution: Refer to the radio's manual for detailed programming instructions. Consider using programming software to manage settings more efficiently. Join forums or local ham radio clubs for advice and assistance.

Preventative Maintenance Tips
- Regularly check and maintain your antenna system.
- Keep the radio and all accessories clean and dry.
- Update the radio's firmware to the latest version.
- Regularly check all connections and cables for wear or damage.

By understanding these common issues and their resolutions, ham radio operators can enjoy a smoother and more reliable experience. Remember, the key to successful ham radio operation lies in regular maintenance, proper setup, and ongoing learning.

Upgrading Your Equipment

Upgrading your equipment, along with maintenance and troubleshooting, is a vital part of ensuring that your ham radio setup remains in optimal working condition. Whether you're a seasoned amateur radio operator or new to the hobby, understanding how to enhance and maintain your equipment is key to enjoying successful, clear communications. Here's a comprehensive guide to help you navigate through upgrading, maintaining, and troubleshooting your ham radio equipment.

Upgrading Your Ham Radio Equipment

1. Assessing Your Needs: Before considering an upgrade, assess your current setup and identify any limitations you're experiencing. Are you looking for extended range, improved audio quality, or additional features such as digital modes or GPS?

2. Research: Once you've identified your needs, research the latest equipment that can offer the improvements you're seeking. Look for product reviews and seek advice from experienced ham operators in forums or local clubs.

3. Compatibility: Ensure that new equipment is compatible with your existing setup. This includes checking for connector types, power requirements, and any additional accessories you might need.

4. Budget: Upgrades can vary significantly in cost. Set a realistic budget that allows for the purchase of quality equipment while avoiding unnecessary financial strain.

5. Future-Proofing: Consider equipment that not only meets your current needs but also has the capability to support future advancements in ham radio technology.

Maintaining Your Ham Radio Equipment

1. Regular Inspections: Conduct regular visual inspections of your equipment for any signs of wear and tear. Check cables, connectors, and antennas for any damage or loose connections.

2. Cleanliness: Keep your equipment clean from dust and debris, which can affect performance. Use a soft, dry cloth to gently wipe down surfaces and compressed air to remove dust from hard-to-reach areas.

3. Software Updates: For digital modes and modern ham radios with computer interfaces, keep your software up to date. Manufacturers often release updates that improve performance or add new features.

4. Preventive Maintenance: Schedule regular preventive maintenance sessions to test all functions of your equipment.

This includes checking the SWR (Standing Wave Ratio) on your antennas and ensuring power supplies are delivering the correct voltage.

Troubleshooting Common Issues

1. Poor Transmission Quality: Check for loose connections, inspect the antenna for damage, and ensure the SWR is within acceptable limits. Poor quality can also result from incorrect microphone gain settings.

2. Reception Issues: If experiencing poor reception, verify that your antenna is correctly oriented and free from obstructions. Adjust the squelch setting to improve signal clarity.

3. Equipment Not Powering On: Ensure all power connections are secure and the power source is functioning correctly. Check fuses and circuit breakers for any signs of failure.

4. Interference: Identify the source of interference by turning off electrical devices one at a time. Utilize filters or change frequencies to mitigate unwanted noise.

5. Software Problems: For issues with digital modes or software-controlled functions, check for software updates or reinstall the software. Ensure your computer meets the software's system requirements.

In conclusion, upgrading your ham radio equipment involves careful planning and research to ensure compatibility and meet your communication needs. Regular maintenance is crucial for longevity and performance, while effective troubleshooting can solve most issues that arise. By staying informed and proactive, you can enjoy the vast world of amateur radio communications with minimal interruptions.

Chapter: 9 The Future of Ham Radio

Technological Advances and Trends

The future of Ham Radio is an exciting intersection of tradition and innovation, where technological advances and trends are shaping a new era for amateur radio enthusiasts. As we look forward, it's essential to understand not only how to use a Ham Radio but also how emerging technologies are expanding the capabilities and possibilities of this enduring hobby.

Understanding Ham Radio Basics

Before diving into the technological advancements, a foundational understanding of Ham Radio operation is crucial. At its core, Ham Radio involves using various frequencies and modes to communicate over distances without relying on the internet or cellular networks. Operators must obtain a license, which requires passing an exam that covers technical knowledge, regulations, and operating practices.

Digital Modes and Software-Defined Radio (SDR)

One of the most significant trends in Ham Radio is the shift from analog to digital modes of communication. Digital modes, such as FT8, D-STAR, and JT65, offer increased efficiency and reliability, especially in weak signal conditions. These modes require specific software and interfaces but open up new avenues for global communication without the need for high-power transmissions.

Software-Defined Radio (SDR) is another game-changer, replacing traditional hardware components with software. SDR allows for more flexible and dynamic operation, with the ability to easily switch frequencies, modes, and even update the radio's capabilities with new software. This adaptability makes SDR a powerful tool for Ham Radio operators, enabling them to experiment with cutting-edge communication technologies.

Internet Integration and Remote Operation

The integration of the internet with Ham Radio operations has led to innovative approaches like remote station operation, where operators can control radio equipment located anywhere in the world through the internet. This technology has made Ham Radio more accessible, allowing enthusiasts to participate in the hobby regardless of their physical location or the space constraints of antenna installations.

Moreover, internet-linked repeater systems, such as Echolink and IRLP, have expanded the reach of VHF and UHF communications, enabling operators to connect across continents using handheld radios connected to local repeaters that are internet-enabled.

Space Communication and Satellites

Amateur radio satellites (AMSAT) and involvement in space communication represent the hobby's cutting-edge frontier. Ham Radio operators can now communicate through satellites, some of which are specifically designed and launched by and for the amateur radio community. This allows for remarkable communication distances and the ability to experiment with signals bounced off the moon or even meteor trails.

Preparing for the Future

To effectively engage with the future of Ham Radio, enthusiasts should focus on understanding and leveraging these technological advances. This involves:

- Gaining a solid foundation in digital modes and SDR, including the software and hardware required.
- Exploring internet-linked communication and remote operation to overcome geographical limitations.
- Participating in space communication projects and satellite operation to push the boundaries of what's possible with amateur radio.

In conclusion, the future of Ham Radio lies in the seamless integration of traditional practices with innovative technologies. By embracing digital modes, SDR, internet capabilities, and space communication, Ham Radio operators can continue to enjoy a fulfilling and evolving hobby that connects people across the globe in new and exciting ways.

The Role of Ham Radio in Modern Communications

Ham radio, also known as amateur radio, has been a vital component of global communication for over a century. Despite the advent of modern technology, including the internet, social media, and cellular networks, ham radio has not only persisted but has also evolved, finding its niche in the digital age. This resilience and adaptability highlight the role of ham radio in modern communications, especially when exploring its future and understanding how to use it effectively today.

The Future of Ham Radio in Modern Communications

Emergency and Disaster Response

One of ham radio's most critical roles in modern communication lies in emergency and disaster response. When natural disasters strike, conventional communication infrastructures can be severely disrupted. Ham radio operators, with their ability to transmit over long distances without relying on the internet or cellular networks, become invaluable. Their ability to quickly establish networks for emergency communication has saved lives during earthquakes, hurricanes, and other catastrophes. This aspect ensures the future relevance of ham radio, as operators continuously train and prepare for emergency response scenarios.

Technological Integration and Advancement

Ham radio is embracing digital technology, integrating with modern communications systems in innovative ways. Digital modes like D-STAR, DMR (Digital Mobile Radio), and System Fusion enhance the clarity and range of communication, allowing for more efficient data transmission. These advancements have attracted a new generation of enthusiasts interested in exploring the intersections between traditional radio and digital technology.

Satellite Communications
Amateur radio satellites (CubeSats) are another exciting frontier. Ham operators can now communicate via satellites, expanding their reach beyond the Earth's curvature. This capability opens up new possibilities for international communication and space exploration support, securing a place for ham radio in the future of communications.

Community and Education
Ham radio fosters a unique global community interested in science, technology, engineering, and mathematics (STEM). Many ham radio clubs provide educational resources, workshops, and certification courses, encouraging young people to pursue careers in these fields. This educational aspect, coupled with the camaraderie among operators, ensures the continuous growth and sustainability of the ham radio community.

How to Use a Ham Radio

Using a ham radio effectively requires understanding its operation, adhering to regulations, and engaging with the community.

Getting Licensed

The first step is obtaining a license. In the United States, the Federal Communications Commission (FCC) oversees the licensing process, which involves passing an examination that covers radio theory, operating practices, and legal regulations. There are different license classes, each granting varying levels of access to radio frequencies.

Equipment and Setup

Ham radio equipment ranges from handheld transceivers to sophisticated base stations. Beginners might start with a VHF/UHF handheld radio to communicate on local repeaters, which extend the range of transmissions. Setting up a station requires an antenna, a power source, and the radio unit itself. Experimenting with different antennas and locations can significantly improve communication capabilities.

Operating Practices

Successful communication on ham radio involves more than just technical setup; it also requires understanding operating etiquette. This includes learning how to properly initiate contact, exchange information, and sign off. Operators should be familiar with the phonetic alphabet and Q codes, shorthand codes used by operators to communicate efficiently.

Engaging with the Community
Finally, the essence of ham radio lies in its community. New operators should connect with local clubs and participate in on-air activities like contests and nets (scheduled meetings on specific frequencies). These interactions are invaluable for gaining experience, learning from seasoned operators, and making friends worldwide.

In conclusion, the role of ham radio in modern communications is multifaceted, blending emergency response, technological innovation, educational outreach, and community building. As it continues to evolve alongside digital technologies, ham radio offers a unique and enduring mode of communication, exploration, and camaraderie. Whether for the hobbyist, the emergency responder, or the tech innovator, understanding how to use ham radio effectively is key to unlocking its full potential in today's digital world.

Continuing Your Education in Radio Technology

Continuing your education in radio technology, particularly focusing on the future of ham (amateur) radio, is an exciting journey that blends traditional communication skills with cutting-edge technology. Ham radio has a storied history but is far from obsolete; in fact, it's evolving with technological advancements. Whether you're a seasoned operator or new to the scene, understanding how to use a ham radio effectively is pivotal in navigating its future. Here's a comprehensive guide:

Understanding the Basics

First, grasp the fundamental concepts of ham radio operations. Ham radio is a hobby and a service that allows operators to communicate across the globe without relying on the internet or cellular networks. It's used for emergency communications, to experiment with wireless communications technology, and for private recreation. Learning the basics involves:

- Getting Licensed: In the U.S., the Federal Communications Commission (FCC) requires operators to have a license. There are three levels: Technician, General, and Extra, each offering increasing levels of access to different radio frequencies.
- Understanding Radio Waves: Knowledge of how radio waves propagate, including frequency bands and modes of transmission,

is crucial. This includes understanding the differences between HF, VHF, and UHF bands, and when to use each.

Advancing Technical Skills

- Equipment Operation: Learn to operate various types of radio equipment, including transceivers, antennas, and tuners. Familiarize yourself with digital modes of communication, such as FT8, WSPR, or DMR, which are increasingly popular.
- Digital Interfaces: Many ham radio operators now incorporate computer interfaces with their setups to facilitate digital communication modes. Understanding software-defined radio (SDR) can significantly enhance your capabilities.
- Emergency Preparedness: Ham radio plays a critical role in emergency communications. Engaging in emergency preparedness exercises and joining amateur radio emergency service (ARES) groups can provide valuable experience.

Engaging with the Community

- Join Clubs and Groups: Becoming a member of ham radio clubs and online forums can provide mentorship opportunities, practical advice, and friendship. Clubs often offer classes to help new members get licensed and improve their skills.
- Contesting and Awards: Participating in contests is a fun way to improve your operating skills and learn from others. It also

motivates operators to fine-tune their stations for optimal performance.

Staying Updated with Technology

- Exploring New Technologies: The ham radio community is continually exploring new technologies. For example, mesh networking with ham radio allows for creating wide-area networks independently of traditional infrastructure.
- Innovation and Experimentation: Amateur radio operators have always been at the forefront of experimentation with radio technology. Engaging in DIY projects, building your own equipment, or experimenting with antenna designs can lead to innovative discoveries.

Educational Resources

- Formal Education: Consider courses in electronics, communications engineering, or computer science to deepen your understanding of the technology behind ham radio.
- Online Resources and Courses: There are numerous online resources, including YouTube tutorials, online courses, and webinars dedicated to ham radio technology and operations.

Looking Ahead: The Future of Ham Radio

The future of ham radio lies in seamlessly integrating traditional radio skills with new technologies. From incorporating internet-linked communications modes to experimenting with satellite communications, the possibilities are vast. Operators who continue their education and adapt to new technologies will lead the way in defining the next era of amateur radio.

In summary, continuing your education in radio technology through ham radio is a blend of respecting tradition and embracing innovation. By mastering the basics, advancing technical skills, engaging with the community, and staying abreast of technological advancements, you'll not only contribute to the hobby but also ensure its relevance in the modern world.

Conclusion

Conclusions about how to use a ham radio effectively hinge on a blend of foundational knowledge, practical experience, and ongoing engagement with the evolving landscape of amateur radio. The journey begins with a solid understanding of the principles underlying radio communications, including the importance of obtaining the necessary licensing through the Federal Communications Commission in the United States. This step not only legalizes your participation but also opens up a world of learning and community engagement.

Understanding the technical aspects of radio operation is crucial. This involves getting familiar with the equipment, from transceivers and antennas to tuners and digital interfaces. As technology advances, so does the equipment used in ham radio, making it imperative for operators to stay updated with the latest tools and techniques. This includes not only traditional voice communication but also digital modes that are becoming increasingly popular for their efficiency and reliability.

Practical experience plays a significant role in mastering ham radio use. This comes from actually operating the radio, participating in on-air activities, and engaging in emergency preparedness exercises. Such activities enhance one's ability to communicate effectively across various frequencies and conditions. They also foster a sense of preparedness for emergency situations, a key aspect of ham radio's value to the community.

Community engagement is another pillar of effective ham radio operation. By joining clubs and groups, operators gain access to a wealth of knowledge and experience from fellow enthusiasts. These communities provide support, education, and the opportunity to participate in contests and awards that further refine operating skills.

Education and continuous learning cannot be overstated. As the technology and regulations around ham radio evolve, so must the operator's knowledge. This may involve formal education in related fields, as well as availing oneself of the vast array of online resources, tutorials, and courses dedicated to amateur radio.

Looking to the future, the effective use of ham radio will increasingly involve integrating new technologies with traditional communication methods. Operators who embrace innovation, whether through digital modes, software-defined radio, or even satellite communications, will find themselves at the forefront of the hobby. Experimentation and innovation within the ham radio community have historically driven advancements in radio technology, and this trend is expected to continue.

In conclusion, effectively using a ham radio is not just about mastering the technical operation of the radio itself but also about embracing a continuous learning mindset, engaging with the community, and contributing to the advancement of the technology. It is a journey that combines respect for the rich history of amateur radio with an enthusiasm for its future possibilities, ensuring that this fascinating hobby remains relevant and vibrant in the digital age.

www.ingramcontent.com/pod-product-compliance
Lightning Source LLC
Chambersburg PA
CBHW071057240526
45471CB00016B/1988